Oliver Jansen

Electron acceleration in the bubble regime

Oliver Jansen

Electron acceleration in the bubble regime

Südwestdeutscher Verlag für Hochschulschriften

Impressum / Imprint
Bibliografische Information der Deutschen Nationalbibliothek: Die Deutsche Nationalbibliothek verzeichnet diese Publikation in der Deutschen Nationalbibliografie; detaillierte bibliografische Daten sind im Internet über http://dnb.d-nb.de abrufbar.
Alle in diesem Buch genannten Marken und Produktnamen unterliegen warenzeichen-, marken- oder patentrechtlichem Schutz bzw. sind Warenzeichen oder eingetragene Warenzeichen der jeweiligen Inhaber. Die Wiedergabe von Marken, Produktnamen, Gebrauchsnamen, Handelsnamen, Warenbezeichnungen u.s.w. in diesem Werk berechtigt auch ohne besondere Kennzeichnung nicht zu der Annahme, dass solche Namen im Sinne der Warenzeichen- und Markenschutzgesetzgebung als frei zu betrachten wären und daher von jedermann benutzt werden dürften.

Bibliographic information published by the Deutsche Nationalbibliothek: The Deutsche Nationalbibliothek lists this publication in the Deutsche Nationalbibliografie; detailed bibliographic data are available in the Internet at http://dnb.d-nb.de.
Any brand names and product names mentioned in this book are subject to trademark, brand or patent protection and are trademarks or registered trademarks of their respective holders. The use of brand names, product names, common names, trade names, product descriptions etc. even without a particular marking in this works is in no way to be construed to mean that such names may be regarded as unrestricted in respect of trademark and brand protection legislation and could thus be used by anyone.

Coverbild / Cover image: www.ingimage.com

Verlag / Publisher:
Südwestdeutscher Verlag für Hochschulschriften
ist ein Imprint der / is a trademark of
OmniScriptum GmbH & Co. KG
Heinrich-Böcking-Str. 6-8, 66121 Saarbrücken, Deutschland / Germany
Email: info@svh-verlag.de

Herstellung: siehe letzte Seite /
Printed at: see last page
ISBN: 978-3-8381-3679-0

Zugl. / Approved by: Düsseldorf, Heinrich-Heine-Universität, Diss., 2014

Copyright © 2014 OmniScriptum GmbH & Co. KG
Alle Rechte vorbehalten. / All rights reserved. Saarbrücken 2014

Contents

Abstract		**3**
Zusammenfassung		**5**
1 Introduction		**7**
2 Theory		**11**
2.1	Plasma	11
	2.1.1 General properties	11
	2.1.2 Defining parameters	12
2.2	Laser-plasma interaction	13
	2.2.1 Non-relativistic regime	13
	2.2.2 Relativistic transparency and self-focussing effect	15
2.3	Synchrotron radiation	16
	2.3.1 Undulators and free electron lasers	17
	2.3.2 Radiation from undulators	19
	2.3.3 Polarisation	20
3 Particle accelerators		**23**
3.1	Conventional accelerators	23
3.2	Laser driven plasma-based accelerators	24
	3.2.1 Linear laser-wakefields	24
	3.2.2 Non-linear wakefields and wave-breaking	26
	3.2.3 Limiting factors and energy gain	28
3.3	Bubble regime	30
	3.3.1 Motivation	30
	3.3.2 Electron movement inside the bubble	34
	3.3.3 Scaling laws for bubble acceleration	35
4 Particle-In-Cell simulation		**39**
4.1	Introduction	39
4.2	Algorithm	42
4.3	Numerical problems and dispersion free solver	44

	4.4	Parallel computing	45
	4.5	Virtual-Laser-Plasma-Laboratory code	47

5 Simulations on bubble acceleration — 49
5.1 Comparison of scaling laws — 49
- 5.1.1 Focal spot size — 51
- 5.1.2 Energy gain and trapped particles — 54
- 5.1.3 Radiation reaction — 57
- 5.1.4 Summary — 64

5.2 Comparison with bubble acceleration experiment — 66
- 5.2.1 Pulse development — 66
- 5.2.2 Electron acceleration — 68
- 5.2.3 Summary — 73

6 Synchrotron radiation and application — 75
6.1 Introduction — 75
- 6.1.1 Calculation and representation of radiation and polarisation — 75
- 6.1.2 Velocity dependent synchrotron radiation — 77
- 6.1.3 Polarisation of synchrotron radiation — 81
- 6.1.4 Summary — 82

7 Towards advanced bubble acceleration schemes — 85
7.1 Introduction — 85
- 7.1.1 Side injection — 85
- 7.1.2 Staging acceleration — 89
- 7.1.3 Summary — 92

8 Summary — 97

9 Outlook — 99

Bibliography — 101

A Acknowledgement — 105

Abstract

The bubble regime of laser-wakefield acceleration has been studied over the recent years as an important alternative to classical accelerators. Several models and theories have been published, in particular a theory which provides scaling laws for acceleration parameters such as energy gain and acceleration length. This thesis deals with numerical simulations within the bubble regime, their comparison to these scaling laws and data obtained from experiments, as well as some specific phenomenona. With a comparison of the scaling laws with numerical results a parameter scan was able to show a large parameter space in which simulation and theory agree. An investigation of the limits of this parameter space revealed boundaries to other regimes, especially at very high ($a_0 > 100$) and very low laser amplitudes ($a_0 \leq 4$). Comparing simulation data with data from experiments concerning laser pulse development and electron energies, it was found that experimental results can be adequately reproduced using the Virtual-Laser-Plasma-Laboratory code.

In collaboration with the Institut für Optik und Quantenelektronik at the Friedrich-Schiller University Jena synchrotron radiation emitted from the inside of the bubble was investigated. A simulation of the movement of the electrons inside the bubble together with time dependent histograms of the emitted radiation helped to prove that the majority of radiation created during a bubble acceleration originates from the inside of the bubble. This radiation can be used to diagnose the amplitude of oscillation of the trapped electrons. During a further study it was proven that the polarisation of synchrotron radiation from a bubble contains information about the exact oscillation direction. This oscillation was successfully controlled by using either a laser pulse with a tilted pulse front or an asymmetric laser pulse.

First results of ongoing studies concerning injecting electrons into an existing bubble and a scheme called 'staging' are presented. The staging scheme utilises the results from injecting electrons by transferring electrons from a bubble acceleration into another bubble. In this manner electrons can be accelerated beyond the usual limitations on the acceleration length of the bubble regime.

Zusammenfassung

Das Bubble Regime der Laser-Kielwellenbeschleunigung wird seit Jahren untersucht auf sein Potential, als Alternative zu klassischen Teilchenbeschleuniger zu dienen. Verschiedene Modelle für das Bubble Regime wurden veröffentlicht, im Besonderen eine Theorie, welche Skalierungsgesetze für Beschleunigungsgrößen beinhalten, wie zum Beispiel der gewonnenen Energie oder der Beschleunigungslänge. Die vorliegende Arbeit beschäftigt sich mit numerischen Simulationen bezüglich des Bubble Regimes, dem Vergleich der Simulationen mit besagten Skalierungsgesetzen und Experimenten, ebenso wie ausgewählten Phänomenen. Eine Untersuchung der Skalierungsgesetze mithilfe numerischer Simulationen ergab einen ausgedehnten Parameterraum, in welchem Simulation und Theorie übereinstimmen. Die Grenzen dieses Parameterraums werden durch Übergänge in andere Regimes definiert, speziell für sehr hohe ($a_0 > 100$) oder sehr niedrige ($a_0 \leq 4$) Laser-Amplituden.

Ein Vergleich von Simulationsergebnissen mit Daten von experimentellen Messungen mit Bezug auf die Entwicklung des Laserpulses und auf Elektronenenergien zeigte, dass experimentelle Resultate mit dem Virtual-Laser-Plasma-Laboratory-Code (VLPL) adäquat reproduziert werden können.

Eine Zusammenarbeit mit dem Institut für Optik und Quantenelektronik der Friedrich-Schiller Universität Jena beschäftigte sich mit der Untersuchung von Synchrotronstrahlung, welche vom Inneren der Bubble emittiert wird. Simulationen der Bewegung der Elektronen innerhalb der Bubble zusammen mit Histogrammen der emittierten Strahlung halfen zu beweisen, dass der Großteil der Strahlung, welche bei einer Bubble Beschleunigung entsteht, aus dem Inneren der Bubble emittiert wird. Diese Strahlung kann zur Diagnose der Schwingungsamplitude der gefangenen Elektronen genutzt werden. In einer weiteren Untersuchung konnte gezeigt werden, dass die Polarisationsrichtung der Strahlung mit der Oszillationsrichtung der Elektronen stark korreliert ist. Diese Oszillationsrichtung wurde erfolgreich beeinflusst mithilfe von asymmetrischen Laserpulsen und Pulsen mit gekippten Pulsfronten.

Des Weiteren werden erste Ergebnisse einer Studie über Elektroneninjektion in existierende Bubbles und das Staging-Verfahren vorgestellt. Beim Staging werden die Ergebnisse der Elektroneninjektion genutzt, um Elektronen aus einer Bubblebeschleunigung in eine neue Bubble zu injizieren. Das Ziel dieses Verfahren ist es, Elektronen jenseits der üblichen Beschleunigungslängen des Bubble Regimes zu beschleunigen.

Chapter 1

Introduction

Modern physics has both a wide range of applications and has helped us understand fundamental concepts of the world we live in. Solid matter physics has taught us how to create advanced materials with incredible attributes for such tasks as withstanding extraordinary conditions or being extremely light, robust and versatile. Thanks to this we also learned much about how matter is structured, how its components interact and what defines its characteristics. Atomic physics has not only made it possible fuse atoms, but also provides a source of knowledge about the very nature of matter on experimentally accessible level.

One field in this wide range of physical disciplines is the field of plasma physics. Plasmas are ionized gases with a large number of individual particles acting in a collective way. Similar to the already mentioned topics it not only has wide ranging applications, including light sources, material processing and potentially even energy generation in the near future. Plasma physics also helps us to gain insight into quite different aspects of nature. This is not very surprising, given the fact, that more than 99% of our solar system and indeed all of the visible universe is in a plasma state. Plasmas include every gaseous form of quasi-neutral matter whose inner dynamics are determined mostly by its own electro-magnetic fields and temperature and which contain enough particles for collective behaviour. Therefore, it is easy to see that quite different types of matter fall under this definition. Among the well known types of plasma are stars, the Aurora Borealis and fluorescent lamps. However, also other types of matter and physical phenomenon can be described as plasmas. This includes intergalactic gases and certain particle clouds like the dust on the surface of the moon. Both are compared to many common types of plasma rather cold and have a low density. Still, most of their behaviour can be modelled as plasma. Also, more abstract phenomenon like quark-gluon plasmas or even the movement of electrons inside an electric conductor do have their place in plasma physics.

A subtopic of plasma physics with very useful applications and an outlook on fundamental research of interest is the field of laser-plasma interaction. While the outcome of applying strong electric fields to a gas whose behaviour is defined by the Lorentz force appears obvious, reality turns out to be much more complicated. In the presence of fields with high intensities, for instance, particles behave in a relativistic fashion. At even higher intensities matter can even be created from vacuum during electron-position pair creation. Also, plasmas in a typical experiment can easily exceed densities of 10^{18} particles per cubic centimetre. However, even though the computational effort in certain problems seems to be immense, much can be learned about the behaviour of astronomic objects, quantum fluctuations, our sun and many more things. In addition to this very fundamental research, cutting edge technology is being developed on the basis of laser-plasma-interactions. These applications include using laser as a cutting tool, diagnostics for fusion plasmas and in recent years particle accelerators.

The demand for particle accelerators has grown over the years. In several fields of physics high energy particles became more and more important. For example, accelerators are sources of fast particles for collision experiments like the ones in CERN, for spectroscopy or as sources for hard X-rays. Furthermore, in medicine sources of fast ions are investigated in cancer therapy. In biology, chemistry and medicine hard X-rays are extremely valuable for their ability in imaging very small structures as for example living cells or even molecules. Such radiation beyond a certain frequency usually is generated in undulators. In undulators fast electrons are forced to do oscillations which causes them to emit cyclotron radiation. The higher the velocity of the electrons the higher is the frequency of the emitted photons as is explained in chapter 2.3. In case of ultra-relativistic electrons undulators can emit high energetic, strongly collimated synchrotron radiation. If the electrons going through an undulator have the same velocity, the emitted photons are coherent and undulator becomes a free-electron-laser (FEL). Such a high quality, high energy beam would lead to fantastic applications as an imaging tool in biology, chemistry and medicine. It also would be a diagnostic tool of enormous value in physics. However, even without the high quality of radiation from a FEL, the undulator still is a very important source of radiation vital for several applications and for research. Undulators and FELs as an example among many applications in dire need of fast electrons require accelerators able to create electron beams suited to emit the radiation necessary for the given task.

Over the last years the requirements on accelerators have grown. These requirements include that the accelerators have to deliver higher energies, mono-energetic particles and just the requirement to be inexpensive. This led to both improvement of existing accelerator concepts as well as to the creation of whole new ones. A concept for plasma based accelerator was conceived by Tajima and Dawson in [1979]. In this concept called laser wake-field accelerator a laser pulse creates a plasma wave. The strong electric gradients generated by this wave can accelerate electrons inserted into the wave. The first accelerators of this type were able to accelerate electrons to Energies up to about $100 MeV$ with a large energy spread. However, they were the basis for more advanced schemes making use of more powerful lasers after the invention of Chirped-Pulse Amplification by Strickland and Mourou in [1985, July]. The main benefits of using such plasma based accelerators is the fact that unlike in conventional accelerators electric fields are not limited by the ability of the walls of the accelerator to withstand them. In common densities for wake field accelerators fields become so strong that accelerations comparable to facility size conventional accelerators can be achieved in plasma accelerators on millimetre or centimetre lengths. One of the more recent developments is a laser-Plasma accelerator utilising the bubble regime, found by Pukhov et al. in [2004, Jul]. The bubble regime can be used as a concept for a very special kind of laser-wakefield accelerator. In such a bubble accelerator electrons travel in the wake field of an ultra-relativistic laser pulse to energies of several $100 MeV$ or even to GeV. This method became popular, because the acceleration could take place over a acceleration distance of several milli- to centimetres. The electrons usually are accelerated in a quasi mono-energetic fashion and the cost of such an accelerator is much lower by orders of magnitude than for instance for a linear or circular accelerators of similar capacities.

Exploring the possibilities of laser-plasma interaction and gaining understanding of plasma physics, not only as a tool for acceleration, requires to understand the physics of a system with a huge number of classical particles. In the field of laser wake-field acceleration common electron densities of 10^{19} and more per cubic centimetre can be encountered. Furthermore the Maxwell's equations, which define the electro-magnetic fields inside a plasma, in the relativistic regime are partial differential equations. This means that not only a large number of problems cannot be solved analytically, but also that numerical simulations are challenging. Several different methods exist in order to face this challenge. One common solution is treat the plasma as a fluid and only consider macroscopic, average values. However, in this specific model kinetic effects, which rely on specific particle movements, are lost. Therefore, another scheme has been developed. The Particle-in-cell (PIC) simulations originate from the works of Buneman and Dawson in [1959] and [1962]. This method is based on stochastically sampling a plasma using macro-particles. These macro-particles correspond to clouds of real particles and reduce the computational effort by orders of magnitude. This way a problem, which would for instance involve 10^{20} particles, can be solved by computing a simulation with about 10^6 or 10^7 particles. These simulation became a vital link between theoretical models and experiments and practical applications. Any theoretical prediction about laser-plasma interaction can be tested in a 'virtual' experiment before a real and probably very expensive and time consuming experiment is conducted. Even though the PIC-method is a very powerful tool, certain problems still require more computational power than common PCs can offer. Therefore computer clusters are necessary, in which several CPU or GPU are joined into a network of parallel working machines.

For these reasons this work deals with the comparison of a well known theory about the bubble regime with particle-in-cell simulations. The theory and its predictions about important scaling laws were published by Pukhov et al. in [2006, Oct]. These scaling laws include predictions about the energy gained from a bubble experiment, the number of particles accelerated and optimal set-up parameters. These predictions were derived using dimensional analysis, which means that dimensionless constants in each scaling law have to be derived numerically. For this work I used the three-dimensional PIC-code 'Virtual laser-plasma Laboratory' (VLPL) developed by my advisor Prof. Alexander Pukhov in [1999]. I chose a laser pulse profile close to common laser profiles in experiments in order to gain the best correlation with experimental studies. The PIC simulations were done in order to find the parameter sub-space in which the scaling laws agree with a quiet realistic simulation of a real experiment. A large parameter space could be found in which theory and simulation matched. This space is limited by transitions into other regimes. At high laser amplitudes and therefore high electron energies synchrotron radiation becomes an important factor. Very small laser pulses as well as very low laser amplitudes seem to be unable to create a stable bubble. Also, with the simulation results I tried to find the dimensionless constants in order to make accurate predictions for experimental applications.

In this work I will present several collaboration with experimental physicists. One of this collaborations dealt with aspects of the movement of electrons during the acceleration, since one major application of accelerated electrons is the generation of radiation. In fact, electrons inside the bubble emit radiation similar to the radiation from undulators. The first collaboration resulted in amongst others the finding, that the majority of radiation, created during a bubble acceleration, is in fact created by trapped electrons. This radiation carries information about the oscillation amplitude of the trapped electrons. The second collaboration did research on the polarisation of the emitted radiation. The polarisation depends on the direction, in which the trapped electrons oscillate. The oscillation direction was successfully controlled by using tilted pulses and asymmetric ones in order to drive the bubble. During my work I needed to make minor changes on the existing VLPL-code which will be discussed in the respective chapters.

Chapter 2

Theory

Before I explain my results in detail, I would like to introduce the main theoretical concepts, which are of interest in my work. Most of the information presented here can be found in text books or review papers like the books by Longmire in [1963], by Chen in [1977] Landau and Lifshitz in [1980] and the review papers by Esarey in [1996] and [2009, Aug] and by Pukhov in [2002, Dec]. I chose to use the *cgs* unit system in the following chapters.

2.1 Plasma

Since the bubble regime is sub-topic of plasma physics, the physical term plasma has to be defined first. This work only deals with plasma in the presence of strong electric fields. Therefore I always assume an ideal, classical, relativistic plasma. Furthermore, since single particle effects will be relevant, a kinetic model will be used. Fluid dynamics and in fact most statistical concepts are insufficient to describe the kinetic effects, necessary for the majority of this work, and therefore will not discussed.

2.1.1 General properties

A plasma is a partially or fully ionised gas. It is quasi-neutral and behaves collectively because of its electromagnetic interactions. 'Quasi-neutral' means, that the plasma as a whole includes as much negatively charged particles as it includes positively charged ones. The collective behaviour comes from the fact that a plasma is dominated by the electromagnetic fields created by the charged particles. In general an ionised gas is considered a plasma if its plasma parameter N_D is very large ($N_D \gg 1$). The plasma parameter of a gas is defined as

$$N_D = \frac{4}{3}\pi \lambda_D^3 n. \tag{2.1}$$

In this equation n is the particle density of the gas, while λ_D is the Debye-length given by

$$\frac{1}{\lambda_D^2} = 4\pi e^2 n \left(\frac{1}{T_i} + \frac{1}{T_e}\right), \qquad (2.2)$$

where T_i is the ion temperature inside the gas and T_e is the temperature of the electrons. The density n is both the density of electrons n_e and ions n_i, since due to the quasi neutrality both densities are equal $n_i = n_e = n$.

In general plasmas can consists of partially ionised gas with particles colliding with each others. Also, the condition $N_D > 1$ is not a condition as strict as the condition for a phase transition in other forms of matter. Therefore certain ionised gases can not be clearly classified as being plasma or not being plasma. In the following work, we always assume a fully ionised gas with $N_D \gg 1$, which only interacts with itself via its electromagnet fields. Therefore the dynamics are mostly defined by Maxwell's equations:

$$\partial_t \mathbf{E} = c\nabla \times \mathbf{B} - 4\pi \mathbf{j} \qquad (2.3)$$

$$\partial_t \mathbf{B} = -c\nabla \times \mathbf{E} \qquad (2.4)$$

$$\nabla \cdot \mathbf{E} = 4\pi \rho \qquad (2.5)$$

$$\nabla \cdot \mathbf{B} = 0 \qquad (2.6)$$

2.1.2 Defining parameters

In order to describe a plasma quantitatively, several parameters come in handy. Since in this work only fully ionised gas is used, the first parameters, which come to mind are the number density and mass of the ions n_I and m_I, their charge q_I and the same quantities of the electrons n_e, $m_e \approx 9.11 \cdot 10^{-28}$g and $q_e = e \approx 4.8 \cdot 10^{-10} \sqrt{\text{g cm}^3/\text{s}^2}$. An other obvious quantity would be the temperature of the plasma given by the average kinetic energy of a particle specie $\langle E_{kin} \rangle = 3/2 k_b T_k$ with k_b being the Boltzmann-constant and k either i or e. In case of plasmas in thermal equilibrium $T_i = T_e$.

From this, other important quantities can be calculated. Disturbances in the particle distribution lead to displacement of charged particles, which in turn locally destroys the quasi-neutrality of the plasma. This leads to electric fields which act on particles against the disturbance. A single particle displaced from its position would oscillate around its original position due to these fields. If a particle cloud of density n_q with charge q would be displaced by a distance r they would feel an electric field corresponding to

$$E = 4\pi n_q q r. \qquad (2.7)$$

For electrons this leads to the classical equation of motion

$$m_e \frac{d^2 r}{dt^2} = -eE. \qquad (2.8)$$

2.2. LASER-PLASMA INTERACTION

From this we get the frequency of the oscillation, which is called the *plasma frequency*

$$\omega_p = \sqrt{\frac{4\pi n_e e^2}{m_e}}. \tag{2.9}$$

Please note, that this only holds in the classical case. If the electrons move with relativistic velocities, m_e becomes $m_e(v) = m_e \gamma$ and equation (2.9) changes to

$$\omega_{p,\text{relativistic}}^2 = \frac{\omega_p^2}{\gamma}, \tag{2.10}$$

with $\gamma = 1/\sqrt{1-v_e^2/c^2}$ the relativistic gamma-factor for the electrons. With the plasma frequency one can derive the length over which electric fields are screened out by electron redistribution. This length is called the *plasma wavelength*

$$\lambda_p = \frac{2\pi c}{\omega_p}, \tag{2.11}$$

with the speed of light $c \approx 3 \cdot 10^{10} cm/s$. An alternative value is the plasma wave number

$$k_p = \frac{2\pi}{\lambda_p} = \frac{\omega_p}{c}. \tag{2.12}$$

A more detailed derivation of this quantities can be found in the literature for instance in the book by Chen from [1977] or the book by Longmire from [1963].

2.2 Laser-plasma interaction

A detailed introduction in laser-physics does not seem to be necessary in order to understand this work. However, the interaction of a laser pulses with a plasma is an important aspect of laser wake-field acceleration. Therefore this chapter contains the most important information and equations regarding laser-plasma interactions.

2.2.1 Non-relativistic regime

In order to accelerate particles, energy must be delivered. In this work the energy always is inserted into the system using a laser pulse. Without going to much into the details of lasers, I just like to present the facts we need for further understanding this work.

A laser pulse is described by its focal spot size R and its duration τ. Furthermore since we usually utilise a circular polarised laser pulse in this work another parameter is the angular frequency $\omega_0 = 2\pi c/\lambda$. This gives the equation for the electric field of a specific laser pulse which travels in positive x-direction as

$$\vec{E} = E_0 e^{-r^2/R^2} \cos\left(\frac{\pi t}{2\tau}\right) \mathrm{Re}\left(e^{k_x x - i\omega t}(\hat{e}_y + i\hat{e}_z)\right), \qquad (2.13)$$

with the amplitude of the electric field E_0 and the wave-number k_x for the propagation direction x. This specific laser pulse follows a Gaußian profile in transversal and a cosine profile in longitudinal direction. If not otherwise stated such a pulse is used in all further calculations and simulations.

Another amplitude, often used to describe a laser pulse is the *normalised, relativistic amplitude* a_0 given by

$$a_0 = \frac{eA_0}{m_e c^2}, \qquad (2.14)$$

with the amplitude A_0 of the vector potential of the electric laser field. This dimensionless parameter is often used, when the laser pulse is strong enough to accelerate electrons to relativistic energies.

For the propagation of a laser pulse inside a plasma, the dispersion relation offers a lot of information. For a light wave in vacuum the dependency of the angular frequency ω on wave number k is

$$\omega = ck. \qquad (2.15)$$

Inside a plasma this equation changes to

$$\omega^2 = (ck)^2 + \omega_p^2. \qquad (2.16)$$

For a laser pulse travelling through a plasma with group velocity v_g and phase velocity v_{ph} it holds that

$$c^2 = v_g v_{ph}. \qquad (2.17)$$

Together with the dispersion relation $\omega = v_{ph} k$ follows

$$k = \frac{v_{ph}\omega}{c^2}. \qquad (2.18)$$

Inserted into eq. (2.16) we receive

$$v_g = c\sqrt{1 - \frac{\omega_p^2}{\omega^2}}. \qquad (2.19)$$

With this, the refraction index η can be derived as

$$\eta = \frac{c}{v_{ph}} = \sqrt{1 - \frac{\omega_p^2}{\omega^2}} = \sqrt{1 - \frac{n_e}{n_{cr}}}. \qquad (2.20)$$

Since η becomes imaginary when $\omega < \omega_p$ only light with higher frequency than the plasma frequency can travel through the plasma. The density at which $\omega = \omega_p$ is called the *critical density* for a certain laser and is given by

2.2. LASER-PLASMA INTERACTION

$$n_{crit} = \frac{\omega^2 m_e}{4\pi e^2}. \tag{2.21}$$

2.2.2 Relativistic transparency and self-focussing effect

Since in the relativistic case the plasma frequency changes according to eq. (2.10) the refraction index changes to

$$\eta = \sqrt{1 - \frac{\omega_p^2}{\omega^2 \gamma}}. \tag{2.22}$$

Equation (2.22) shows that laser pulses with $\omega < \omega_p$ can enter the plasma as long as the electrons in front of the pulse are fast enough.
Since in the relativistic regime the mass of an electron is a function of its velocity $m_e = m_e(v)$, we have to use eq. (2.10) in order to calculate the refraction index. This means the relativistic group velocity increases to

$$v_{g,rel} = c\sqrt{1 - \frac{\omega_{p,rel}^2}{\omega^2}} = c\sqrt{1 - \frac{\omega_{p,rel}^2}{\omega^2 \gamma}}, \tag{2.23}$$

while the relativistic refraction index increases to

$$\eta_{rel} = \sqrt{1 - \frac{n_e}{n_{cr} \gamma}}. \tag{2.24}$$

This increase of the refraction index acts like a collecting lens and acts against the usual defocussing of a laser pulse behind its focus. So with the correct relativistic plasma frequency one can send a laser pulse over long distances through a plasma without suffering from defocussing. Esarey et al. summarised in [1996] that in the limit $a^2 \ll 1$ the development of the radius R of a laser pulse can be approximated as

$$\frac{dR}{dz^2} = \frac{1}{Z_R R^3}\left(1 - \frac{P}{P_c}\right), \tag{2.25}$$

with the Rayleigh length

$$Z_R = \frac{\pi R^2}{\lambda}, \tag{2.26}$$

the normalised spot radius $R = r/\lambda$ and $P/P_c = k_p^2 a_0^2 r^2 / 16$. The solution to the differential equation (eq: RdevDiff) is

$$\frac{r}{r_0} = 1 + \left(1 - \frac{P}{P_c}\right)\frac{z^2}{Z_R^2}, \tag{2.27}$$

with r_0 being the initial spot size of the laser pulse. Equation (2.27) states that the closer the pulse power P gets towards the critical power P_c the less the laser pulse suffers from diffraction. Equation (2.25) and therefore (2.27) have been derived for $a_0 \ll 1$ which led to the approximation $(1 + a_0^2)^{-1/2} \approx 1 - a_0^2/2$ and with this to the prediction of over-focussing for $P > P_c$. This is not a physical behaviour as Sprangle et al. showed in [1987]. In fact for large powers $P > P_c$ and especially for large laser pulses $R > \lambda_p$ it can be shown and observed in simulations that a laser pulse has a width that oscillates around a stable spot size. The theory by Sprangle et al. states that pulses small enough

$$L \leq \frac{\lambda_p}{1 + a^2} \quad (2.28)$$

do not benefit from this relativistic self-focussing and that they only observed this behaviour typically for

$$L > \lambda_p. \quad (2.29)$$

However, as the findings presented in chapter (5.1) suggest, that equation (2.29) seems to be too strict.

2.3 Synchrotron radiation

Since the main focus of this work is fast, charged particles in strong fields, it is necessary to consider radiation emitted by these particles due to acceleration. This leads to synchrotron radiation in case of electrons trapped and accelerated inside a bubble. During their acceleration inside the bubble, electrons oscillate around their propagation direction. This is because the particles do have a transversal momentum while entering the bubble and because of the transversal electric field inside the bubble. In the review by Esarey et al. from [2002, May] it is shown that the length of such a betatron oscillation can be calculated as

$$\lambda_\beta = \pi r_b \sqrt{\frac{2\gamma}{\Phi_0}} \quad (2.30)$$

with bubble radius r_b and the normalised wake-amplitude Φ_0, which can be calculated for the bubble regime as

$$\Phi_0 \approx \frac{k_p^2 r_b^2}{4}, \quad (2.31)$$

with $k_p = \omega_p/c$. With this it follows, that

$$\lambda_\beta = \frac{2\pi c}{\omega_p} \sqrt{2\gamma} \quad (2.32)$$

The resulting betatron frequency is

$$\omega_\beta = \frac{\omega_p}{\sqrt{2\gamma}}. \quad (2.33)$$

2.3. SYNCHROTRON RADIATION

Wit this it is possible to calculate the radiation emitted by electrons travelling through the magnetic fields $\vec{B}$ of a bubble. The wave length of the synchrotron radiation is

$$\lambda_s = \frac{\lambda_\beta}{2\gamma^2} = \frac{\sqrt{2}\pi c}{\omega_p}\gamma^{-3/2}. \tag{2.34}$$

The energy of this radiation equals the energy lost by the emitting particle. While at low particle velocities this loss of energy is small compared to the overall particle energy, at relativistic velocities the energy radiated can not be neglected any more. The total intensity radiated in every direction can be calculated as

$$I_{synch} = \frac{2e^4 \vec{B}^2 v^2 \gamma^2}{3m^2 c^5} \tag{2.35}$$

as shown for instance by Landau et al. in [1980] §74. It can be calculated, that the radiation for large particle energies is concentrated in the plane of oscillation. Be $\Delta\Phi$ the angular range in which most of the radiation is emitted. It then holds that

$$\Delta\Phi \approx \frac{1}{\gamma}. \tag{2.36}$$

As shown by Landau et al. in [1980] at high energies the radiation emitted within $\Delta\Phi$ inside a uniformly, electric field is linearly polarised in the direction of the field. This information will become quite useful when electron movements inside the bubble are discussed further.

Another reason why synchrotron radiation is important in the field of electron acceleration is the fact, that it is a important source of high energy radiation. X-radiation, radiation with wavelength between 0.01 and 10 nanometres, can be created using the Bremsstrahlung effect. Since the cross section for the Bremsstrahlung effect decreases with the velocity of electrons, for radiation with significantly higher energies other effects are necessary. A method in order to create hard X-rays using fast electrons is to use *undulators*.

2.3.1 Undulators and free electron lasers

An undulator as proposed by Motz et al. in [1950] consists of two rows of magnets (see fig. (2.1) on page 18) through which an electron beam of high velocity is guided. In each row, the polarity of the magnets swaps with each magnet to the opposite polarity. While moving forward, the electrons oscillate through the changing magnetic field. One oscillation is completed after the distance λ_u which is the distance between two magnets of the same polarity. During this oscillations the electrons emit photons almost in forward direction as given by eq. (2.36).

Undulator are very simple, yet very reliable and efficient sources for synchrotron radiation when the electrons used are fast enough and therefore the angular spread of the emitted photons is low.

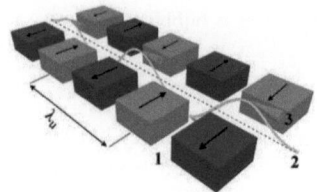

Figure 2.1: *Principle of an undulator. 1 denotes the magnets of opposing polarity, 2 the electron beam and 3 the emitted photons. λ_u periodicity of the magnetic field.*

If now the energies of all the electrons inside of an undulator are more or less the same, the energy of the emitted photons is very similar as well. In case of fast, monoenergetic electrons, the resulting radiation would be monochromatic and coherent. In this case the undulator becomes a *Free Electron Laser* (FEL). High energy synchrotron radiation, especially coherent one, has many application. Amongst these are the imaging of very small objects, since the size of the smallest object one can resolve using light is proportional to the wavelength of that light. In order to make very small objects like molecules directly visible, high energetic radiation is needed. The Abbe-limit

$$d = \frac{\lambda}{2n \sin \Theta} \qquad (2.37)$$

defines the smallest size d light of a wavelength λ can resolve while travelling through a medium with refraction index n while converging to a spot at an angle Θ. $n \sin \Theta$ also is called numerical aperture (NA) and in optics of today the practical limit is at about

$$NA \lesssim 1.4. \qquad (2.38)$$

With this, the smallest object a light source of wavelength λ can image has a size of

$$d \approx \frac{\lambda}{2.8}. \qquad (2.39)$$

Using coherent laser light, it would be possible to create three-dimensional, holographic images of small objects. Also, if the velocity of the electrons used in a FEL is scalable, than the wavelength of the laser light is scalable as well. With reliable sources of high energy electrons and scalable electron energies the wavelength of FELs would be adjustable in a regime where there even are no laser sources at the current time.

2.3. SYNCHROTRON RADIATION

2.3.2 Radiation from undulators

The wavelength of light emitted by an undulator can be calculated as shown by Luchini et al in [1990 L]. In a system co-moving with the electron, the distance λ_u becomes

$$\lambda'_u = \frac{\lambda_u}{\gamma} \tag{2.40}$$

and the electron emits radiation with the frequency

$$f' = \frac{c}{\lambda'_u} = \frac{c\gamma}{\lambda_u}. \tag{2.41}$$

Due to Doppler-shift, this frequency transforms in the laboratory system to

$$f = \frac{f'}{\gamma(1 - \beta \cos\theta)} = \frac{c}{\lambda_u(1 - \beta \cos\theta)}, \tag{2.42}$$

under the angle θ towards the propagation direction. With the Taylor-expansion of cosine $\cos\theta = 1 - \theta^2/2 + ...$ up to the second order and $1 + \beta \approx 2$, which leads to $\gamma^2 \approx 1/(2 - 2\beta)$, one can calculate the observed frequency

$$f = \frac{c}{\lambda_u(1 - \beta \cos\theta)} \approx \frac{2\gamma^2 c}{\lambda_u(1 + \gamma^2 \theta^2)} \tag{2.43}$$

and with it the wavelength

$$\lambda \approx \frac{\lambda_u}{2\gamma^2}(1 + \gamma^2 \theta^2). \tag{2.44}$$

The Lorentz factor of an electron moving through a magnetic field of the shape $\vec{B} = B_0 \cos(2\pi z/\lambda_u)\hat{e}_y$ can be calculated with the equation of motion

$$\frac{d\vec{p}}{dt} = -e(\vec{E} + \vec{v} \times \vec{B}) = -e(\vec{v} \times \vec{B}). \tag{2.45}$$

With $\vec{p} = \gamma m \vec{v}$ this equation gives us in forward direction x

$$m\gamma \frac{dv_x}{dt} = e\frac{dz}{dt} B_0 \cos\left(\frac{2\pi z}{\lambda_u}\right). \tag{2.46}$$

After integrating both sides we receive

$$m\gamma v_x = \frac{eB_0 \lambda_u}{2\pi} \sin\left(\frac{2\pi z}{\lambda_u}\right). \tag{2.47}$$

which transforms into

$$v_x = \frac{Kc}{\gamma} \sin\left(\frac{2\pi z}{\lambda_u}\right). \tag{2.48}$$

with the definition of the magnetic deflection strength

$$K := \frac{eB_0\lambda_u}{2\pi mc}. \qquad (2.49)$$

In case of $K \gg 1$ usually the undulator is called *wiggler*. Since the electron does not change its energy passing through the magnetic field $\gamma = \sqrt{(1 - (v_x^2 + v_z^2)/c^2)}^{-1}$ is a constant and therefore

$$\frac{v_z^2}{c^2} = 1 - \frac{1}{\gamma^2} - \frac{K^2}{\gamma^2}\sin^2\left(\frac{2\pi z}{\lambda_U}\right). \qquad (2.50)$$

This leads to the approximate

$$\frac{v_z}{c} \approx 1 - \frac{1 + \frac{K^2}{2}}{2\gamma^2} + \frac{K^2}{4\gamma^2}\cos\left(\frac{4\pi z}{\lambda_U}\right), \qquad (2.51)$$

which again can be averaged over a full oscillation as

$$\frac{\bar{v}_z}{c} = 1 - \frac{1 + \frac{K^2}{2}}{2\gamma}. \qquad (2.52)$$

Using this result we can define an effective Lorentz factor

$$\gamma_K = \frac{\gamma}{\sqrt{1 + \frac{K^2}{2}}}, \qquad (2.53)$$

which substituted into equation (2.44) gives the final undulator equation

$$\lambda = \frac{\lambda_u}{2\gamma^2}\left(1 + \frac{K^2}{2} + \gamma^2\theta^2\right), \qquad (2.54)$$

for $\beta \approx 1$ and as an Taylor expand for θ to the second order.

2.3.3 Polarisation

Polarisation of the radiation can be an important factor as discussed later in chapter 6.1.3. In general, polarisation of any transversal wave is the direction of the oscillation of that wave. Since the oscillation direction of the magnetic and the electric field of a electromagnetic wave are perpendicular, it is arbitrary which one to discuss. Therefore in the following I choose the electric field polarisation as representation of the polarisation of the whole electro-magnetic field. Polarisation of a single wave can be *linear*, *circular* or *ellipsoid*. A linear polarised wave oscillates in one plane. Projected on a plane perpendicular to the propagation direction, the oscillation takes place on a single line. Circular polarisation can be understood as two oscillations superimposed: A sinusoidal one and a co-sinusoidal one, perpendicular

2.3. SYNCHROTRON RADIATION

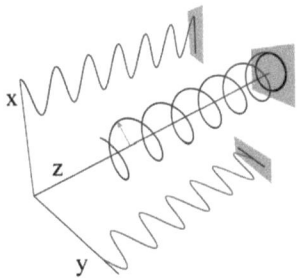

Figure 2.2: *The green arrow marks the polarsation of a wave, propagating in z-direction. It follows the purple path. Blue and red denote the projections of the circular polarisation onto the $x/z-$ and respectively $y/z-$plane. The projections correspond to linear polarisation in the x-direction (blue) and the $y-$direction (red).*

to the first one. The polarisation is called ellipsoid in case, both oscillations are not in phase (and have no phase difference of 90°) or do not have the same amplitude, but have a fixed phase difference and amplitude ratio.

The polarisation depends on the way, a wave was created. For instance, a single particle oscillating in a plane would emit radiation polarised in the same plane. Polarisation can also be changed by reflection or refraction. Most sources of radiation include electromagnetic waves of different polarisation. If the polarisation is uncorrelated, which means that the radiation has a mixture of different kinds and/or directions of polarisation, it is called unpolarised. A clarification of circular polarisation can be seen in figure (2.2) on page 21. In comparison to this, ellipsoidal polarisation is illustrated in figure (2.3) on page 22.

Equation 2.13 on page 14 describes the electric field of a circular polarised pulse. The electric field of a pulse, polarised linearly in $y-$direction, is given by

$$\mathbf{E} = E_0 e^{-r^2/R^2} \cos\left(\frac{\pi t}{2\tau}\right) \cos(\omega_0 t - k_x x)\mathbf{e}_y. \qquad (2.55)$$

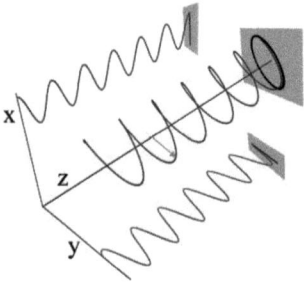

Figure 2.3: *The same as figure (2.2) on page 21, but now with an ellipsoidal polarised wave. The amplitude of the $x-$component (blue) has been reduced, while the $y-$component (red) is slightly out of phase.*

Chapter 3
Particle accelerators

Beams of charged particles are integral to a number of applications. These include generation of high energy radiation, fundamental research on particles, biomedical applications and industrial processing. Two very common accelerators for high energy particles are linear accelerators and cyclotrons. Most particle accelerators use electric fields E to bring particles to higher energies W according to

$$W = qEL, \tag{3.1}$$

where q is the charge of the accelerating particle and L is the length over which it is accelerated. The strength of the accelerating field and the acceleration length can easily become limiting factors.

3.1 Conventional accelerators

The most basic accelerator for particles especially electrons is the Linear Accelerator ($LinAc$). The electric field of a ring-shaped cathode attracts an approaching electron until the electron passes through. Than the cathode is switched into an anode and repels the electron forward. A LinAc consists of several of such *drift tubes* lined up after each other (see figure (3.1) on page 28). Between the drift tubes the electrons gain kinetic energy according to equation (3.1) on page 23. The advantage of a LinAc is the fact that its principle is rather simple and the accelerated particle does not loose energy because of synchrotron radiation (see chapter (2.3)). The downside is the size of a LinAc. As an example the Stanford University Linear Accelerator ($SLAC$) at the time of writing has a length of two kilometres and accelerates electrons to energies of up to $50 GeV$. An significant increase in the electric field in such accelerators is impossible due to dielectric breakdown (usually below $100 MV/m$) of the materials the accelerator consists of. Therefore the length of a LinAc is the only parameter, which can be freely adjusted in order to obtain a specific energy. This of course can lead to very long accelerators.

Because of these dimensions the cyclotron, synchrotron and similar accelerators where invented. Since this work is focused on high energy particles and super-relativistic lasers, synchrotrons are going to be used representing all accelerators with bend particle trajectories. These accelerators force electrons on circular or spiral trajectories in order to achieve long acceleration length on a small area. The downside of this approach is that charged particles on bend trajectories do loose energy by emitting photons (see chapter (2.3)). This leads to the hindering fact that with increasing particle velocity it becomes increasingly difficult to further accelerate that particle. This together with the upper limit for electric fields also gives an upper limit for any given synchrotron-accelerator. This upper limit can be very high especially when the curvature and therefore the energy loss due to synchrotron radiation is very low. Still, this makes is necessary to build very strong and expensive field coils and large facilities. Particle accelerators like linear accelerators or synchrotrons are quite common today, but still only few of them are able to accelerate electrons to more than a few GeV.

3.2 Laser driven plasma-based accelerators

In plasmas stronger electric fields can be achieved due to the high particle density and the fact that ionisation of material is not a problem. Given a perturbation strong enough to move this high densities, fields in the order of $100GV/m$ can be achieved as shown by Gordon et al in [1998]. Such perturbations can be caused by modern lasers. The very concept of the first *Laser-Plasma-Accelerator* (LPA) was proposed by Tajima et al. in [1979]. The first LPA though were only able to accelerate a small number of electrons to energies slightly above $100MeV$ while the majority of accelerated electrons remained relatively slow ($< 10MeV$). There is a range of different LPA concepts starting from using long laser pulses with moderate energy (*Plasma-Beat-Wave-Accelerator*) to high energy, short pulses (*Laser-Wakefield-Accelerator* (LWFA)). With the invention of Chirped-Pulse-Amplification the LWFA and its high relativistic subtype the *Bubble-/Blow-out-/Caviation-Regime-Accelerator* became more and more interesting. The electron acceleration within the bubble regime will be the main object of this work.

3.2.1 Linear laser-wakefields

All LPAs create wakefields due to ponderomotive force in order to accelerate particles. The following derivation can be found in the book by Kruer et al. from [2003(Kr)]. Since the linear regime will not be part of my work I will only summarise the main points briefly. For a particle species j with phase space distribution $f_j(\vec{x}, \vec{v}, t)$ starting from the continuity equation

$$\frac{\partial f_j}{\partial t} + \frac{\partial(\dot{\vec{x}} f_j)}{\partial \vec{x}} + \frac{\partial(\dot{\vec{v}} f_j)}{\partial \vec{v}} = 0 \qquad (3.2)$$

3.2. LASER DRIVEN PLASMA-BASED ACCELERATORS

we obtain the Vlasov equation

$$\frac{\partial f_j}{\partial t} + \vec{v}\cdot\frac{\partial f_j}{\partial \vec{x}} + \frac{q_j}{\gamma m_j}\left(\vec{E}+\frac{\vec{v}\times\vec{B}}{c}\right)\cdot\frac{\partial f_j}{\partial \vec{v}} = 0 \tag{3.3}$$

using

$$\dot{\vec{x}} = \vec{v} \quad \text{and} \quad \dot{\vec{v}} = \frac{q_j}{\gamma m_j}\left(\vec{E}+\frac{\vec{v}\times\vec{B}}{c}\right). \tag{3.4}$$

With the mean velocity $\vec{u}_j$

$$n_j \vec{u}_j = \int \vec{v} f_j(\vec{x},\vec{v},t) d\vec{v} \tag{3.5}$$

the Vlasov equation becomes the force equation

$$n\frac{\partial \vec{u}}{\partial t} + n\vec{u}\frac{\partial \vec{u}}{\partial \vec{x}} = \frac{nq}{\gamma m}\left(\vec{E}+\frac{\vec{u}\times\vec{B}}{c}\right) - \frac{1}{m}\frac{\partial P}{\partial \vec{x}}, \tag{3.6}$$

where P is the pressure. For negligible electron pressure and an electric field, which oscillates fast with the frequency ω, this equation can be averaged over fast, local oscillation of electrons in order to obtain

$$m\frac{\partial \vec{u}}{\partial t} = -e\vec{E} - \frac{e^2}{4m\omega^2}\nabla \vec{E}^2(\vec{x}). \tag{3.7}$$

This equation now defines the ponderomotive force

$$\vec{F}_p = -\frac{e^2}{4m\omega^2}\nabla \vec{E}^2(\vec{x}) \tag{3.8}$$

as a force on particles (here electrons) by a strong, quickly oscillating field, pushing away the particles from regions of high field pressure. It should be kept in mind, that equation (3.8) was calculated for electrons neglecting their pressure and averaging out fast, local movement of electrons.

With equation (3.8), describing the ponderomotive force, linear plasma waves created by a laser pulse can be described analytically. In one dimension even for non-linear waves using the cold fluid approach (the continuity equation, equation (3.6) and the Poisson's equation for electro-statics) the problem can be solved analytically. The non-linear three dimensional case however requires numerical simulations. As can be read in the review by Esarey et al. from [1996] in the linear regime of laser-wakefield acceleration $a_0 \ll 1$ the plasma wave generated is described by

$$\left(\frac{\partial^2}{\partial t^2}+\omega_p^2\right)\frac{n}{n_0} = c^2 \nabla^2 \frac{a^2}{2}, \tag{3.9}$$

where $\vec{a} = e\vec{A}/m_e c^2$ is the normalised potential of the electric vector-potential $\vec{A}$, ω_p is the plasma frequency and n/n_0 is the perturbed density of the plasma wave. The electric field follows the equation

$$\nabla^2 \phi = 4\pi(n_e - n_i), \tag{3.10}$$

where ϕ is the electric potential, n_i the ion density and n_e the electron density. For a circular polarised laser pulse with

$$a^2 = a_0^2 \exp\left(\frac{-2r^2}{r_L^2}\right) \sin^2\left(\frac{\pi\zeta}{L}\right), \tag{3.11}$$

with r_L the width of the laser pulse, L the length and $\zeta = x - ct$ the relative longitudinal coordinate. Behind the laser pulse, which is for $\zeta < 0$ the electric field in forward direction is given by

$$\frac{E_x}{E_0} = -\frac{\pi}{4} a_0^2 \exp\left(-\frac{2r^2}{r_L^2}\right) \cos(k_p \zeta), \tag{3.12}$$

while the plasma disturbance is

$$\frac{n}{n_0} = -\frac{\pi}{4} a_0^2 \left[1 + \frac{8}{k_p^2 r_L^2}\left(1 - \frac{2r^2}{r_L^2}\right)\right] \exp\left(-\frac{2r^2}{r_L^2}\right) \sin(k_p \zeta). \tag{3.13}$$

$E_0 = m_e c \omega_p / e$ here is the non-relativistic wave breaking field and in this regime $E \ll E_0$. The radial force which is a sum of the electric field E_r and the magnetic field $-B_\theta$ can be calculated according to the Panofsky-Wenzel theorem

$$\frac{\partial E_x}{\partial r} = \frac{\partial (E_r - B_\theta)}{\partial \zeta}, \tag{3.14}$$

which gives

$$\frac{E_r - B_\theta}{E_0} = \frac{a_0^2 r \pi}{r_L^2 k_p} \exp\left(-\frac{2r^2}{r_L^2}\right) \sin(k_p \zeta). \tag{3.15}$$

While equation (3.9) and (3.10) are true for all non-relativistic, laser driven, linear wakefields ($a_0 \ll 1$ which implies $E \ll E_0$), the equations (3.12), (3.15) and (3.13) were derived for a specific laser pulse given by equation (3.11).

3.2.2 Non-linear wakefields and wave-breaking

The non-linear regime of wakefield acceleration can be solved in one dimension as has been reviewed by Esarey et al. in [2009, Aug]. With a quasi-static approach using fluid momentum and the continuity equation, the Poisson's equation for such a one-dimensional plasma $\partial^2 \phi / \partial \zeta^2 = k_p^2(n/n_0 - 1)$ can be transformed into

$$\frac{\partial^2 \phi}{\partial \zeta^2} = k_p^2 \gamma_p^2 \left[\beta_p \left(1 - \frac{1+a^2}{\gamma_p^2(1+\phi)^2}\right)^{-\frac{1}{2}} - 1\right], \tag{3.16}$$

3.2. LASER DRIVEN PLASMA-BASED ACCELERATORS

where $\gamma_p = (1-\beta_p^2)^{-1/2} \approx \omega_0/\omega_p$ and $\beta_p = v_p/c$ with the wake field phase velocity v_p. The phase velocity of the wake field can be approximated with the group velocity of the laser pulse creating it

$$v_p = \frac{c}{\sqrt{1 - \frac{\omega_p^2}{\omega_0^2}}}, \qquad (3.17)$$

where ω_0 is the laser frequency. From equation (3.16) the density perturbation

$$\frac{n}{n_0} = \gamma_p^2 \beta_p \left[\left(1 - \frac{1+a^2}{\gamma_p^2(1+\phi)^2}\right)^{-\frac{1}{2}} - \beta_p \right] \qquad (3.18)$$

and the average fluid velocity

$$u_x = \gamma_p^2(1+\phi)\left[\beta_p - \left(1 - \frac{1+a^2}{\gamma_p^2(1+\phi)^2}\right)^{\frac{1}{2}}\right] \qquad (3.19)$$

follow.

From equation (3.16) follows that behind the laser pulse ($a_0 = 0$) the potential has extrema

$$\phi_m = \frac{\hat{E}_{max}^2}{2} \pm \beta_p \sqrt{\left(1 + \frac{\hat{E}_{max}^2}{2}\right)^2 - 1}, \qquad (3.20)$$

with the normalised maximum of the electric field $\hat{E}_{max} = E_{max}/E_0$.

For higher dimensions usually simulations are used. Common methods for simulating non-linear, relativistic laser plasma interactions include non-linear (quasistatic) plasma fluid models as used by Shadwick et al. in [2002, Feb] as well as Particle-In-Cell-simulations as presented in chapter (4). The two dimensional case will not be part of this work, therefore I will not discuss it in detail.
As can be read in the reviews by Esarey et al. from [2009, Aug] or Pukhov et al. from [2002, Dec] the theory of strongly non-linear plasma waves and the wave breaking (see below) of plasma wakefields is complex and can only partially described analytically. In the following I like to consider a pulse as the one given by equation (3.11) with a length $L < \lambda_p/2$. Pulses which are longer than one half of the plasma wave length in the non-linear regime experience strong modulations and lead to a different regime, the so called self-modulated laser-wake field acceleration (as described by Esarey et al. in [1996]). A short laser pulse however creates a wake field, which has a maximum given by

$$\frac{E_{max}}{E_0} = \frac{a_0^2}{\sqrt{1+a_0^2}}. \qquad (3.21)$$

For $E_{max} > E_0$, with $E_0 = cm_e\omega_p/e$ the wake field becomes highly non-linear. Inserting equation (3.21) into (3.20) for the negative sign gives

$$\frac{1}{\gamma_m} = \phi_m + 1. \qquad (3.22)$$

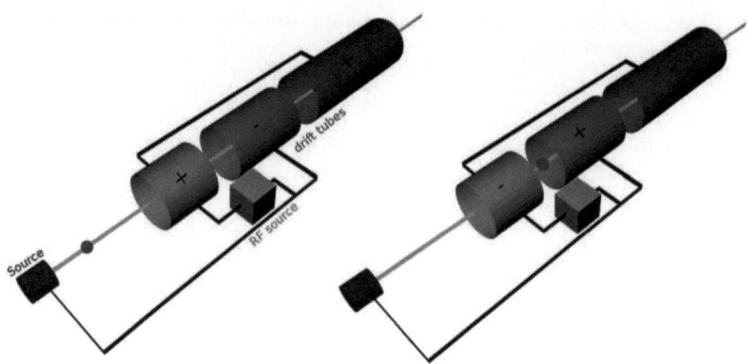

Figure 3.1: Schema of a Linear Accelerator according to Gustav Ising and Rolf Wideröe. The source emits electrons which are attracted by positively charged drift tubes in front of them and repelled by negative ones behind of them. The length of the tubes is chosen so that the electrons take the same time passing through the tube as the RF source in order to change the charge of the tubes.

Equation (3.22) inserted into equation (3.19) leads to $u_x = \beta_p$ (again, behind the laser: $a_0 = 0$), while equation (3.18) diverges with $n \to \infty$. This leads to *wave breaking* of the plasma wave behind the laser pulse when electrons start to move with phase velocity and the periodical structure of the wake is destroyed.
With the assumption of a cold plasma the maximum electric field of a periodic plasma wave, which is the same as the minimum electric field in order to achieve wave breaking, can be estimate as

$$\frac{E_{wb}}{E_0} = \sqrt{2(\gamma_p - 1)}. \qquad (3.23)$$

Wave breaking turned out to be the key mechanism for self-injection of electrons into an accelerating potential just behind the laser pulse, where these electrons can be accelerated. With well adjusted laser parameters this method leads to another acceleration regime which will be the main topic of this work.

3.2.3 Limiting factors and energy gain

In order to obtain the energy gain of an electron in a LWFA one first has to understand which mechanisms limit the acceleration of electrons. One of this mechanisms is the difference in the velocities of the accelerated electrons and the accelerating

3.2. LASER DRIVEN PLASMA-BASED ACCELERATORS

wake. The phase velocity v_p of the plasma wake can be approximated by the group velocity of the laser pulse which according to equation (2.19) always is lower than the speed of light. In the one-dimensional low-intensity limit in fact

$$\gamma_p \approx \frac{\omega}{\omega_p}, \tag{3.24}$$

which is equivalent to

$$\beta_p = \sqrt{1 - \frac{\omega_p}{\omega}} < 1 \tag{3.25}$$

for any under-dense plasma $\omega_p < \omega$, which is equivalent to $v_p < c$. This also means that high energy electrons with velocity $v_x \approx c$ move faster than the plasma wave. This can lead to a situation called *dephasing* in which electrons leave the accelerating phase of the wave, which is the half of one wave period with electric fields accelerating electrons. After leaving this phase, electrons enter the following decelerating phase of that wave period, in which the sign of the electric field changes. Obviously, electrons only gain energy during the accelerating phase. The *dephasing length* L_D is the distance a relativistic electron has to travel in order to cross one half of the plasma wave period. Since the decelerating length grows linearly with increasing quotient v_p/v_x until dephasing would vanish in the limit $v_p \to v_x$, which is $L_D \to \infty$, the dephasing length can be described with

$$\left(1 - \frac{v_p}{v_x}\right) L_D = \frac{\lambda_p}{2}. \tag{3.26}$$

Again for $v_x \approx c$ and with a Taylor expansion of $1 - \beta_p^2$ around the point $\beta_p = 1$ we can write

$$2\left(1 - \frac{v_p}{v_x}\right) \stackrel{v_x \approx c}{=} 2\left(1 - \frac{v_p}{c}\right) \stackrel{FE}{\approx} \frac{1}{\gamma_p^2}. \tag{3.27}$$

With equation (3.27) equation (3.26) becomes

$$L_D \approx \gamma_p^2 \lambda_p. \tag{3.28}$$

With equation (3.1) on page 23 one can estimate the maximum energy W_{max} an electron can gain during its acceleration as

$$W_{max} = eE_{max}L_D \approx 2\pi\gamma_p^2 \frac{E_{max}}{E_0} m_e c^2, \tag{3.29}$$

with the last approximate given by Esarey et al. in [1996].

The second mechanism, which impairs the energy gain of electrons in LWFA and which I would like to discuss briefly, is the depletion and diffraction of the driving laser pulse. While the laser pulse drives a wake field and accelerates particle, it constantly looses energy and it is subject to Rayleigh diffraction. Rayleigh diffraction limits the acceleration length for most LWFA. As it is well known and can be read in textbooks (as for instance the one by Eichler from [1963 E]), the spot size R of a Gaussian pulse propagating in x-direction behaves according to

$$R(x) = R_0\sqrt{1 + \frac{x}{Z_R}}, \qquad (3.30)$$

with the Rayleigh length Z_R of a laser pulse given by equation (2.26). After travelling the distance $x = Z_R$, the pulse widened by a factor of $\sqrt{2}$. In vacuum the diffraction distance L_{diff} equals the Rayleigh length Z_R. In plasmas mechanisms like beam guide lining as described by Esarey et al. in [1996] or the relativistic self-focussing effect (chapter (2.2.2)) can be used, in order to increase the diffraction distance.
The depletion length L_{dp}, which is the distance a laser pulse can travel before using up too much energy in order to drive a wake field in the given regime, is a little bit more difficult to handle. An easy approximate can be done by equating the total laser energy W_L to the energy of the wake field driven by the pulse

$$E_x L_{dp} \approx E_L L = W_L \qquad (3.31)$$

ignoring the transversal electric field. In this equation E_L is the electric field of the laser pulse. Obviously for the most efficient acceleration one has to aim for matching these three lengths $L_D \approx L_{dp} \ll L_{diff}$.

3.3 Bubble regime

3.3.1 Motivation

While the first of the accelerators the LWFA was proposed for non-relativistic laser pulses, *bubble accelerators* operating in the bubble regime only work with laser pulses of relativistic intensities. Such strong laser fields became possible with the invention of the *chirped pulse amplification* by Strickland and Mourou in [1985, July]. At this laser intensities the amplitudes of the wake-field can become high enough for wave-breaking to occur. In case of this scenario the periodic wake behind the laser pulse breaks and a single spherical structure, the so called *bubble*, is formed ([2002, Mar]). This regime can not be described by linear plasma theory, which is the reason why numerical simulations are of utmost importance for a deeper understanding of the bubble regime.
Laser pulses necessary for bubble acceleration are so strong that almost all the plasma electrons are expelled from the volume of the bubble due to the strong ponderomotive force of the laser pulse. Some electrons can be trapped inside the bubble during the wave-breaking. This trapping of background electrons is called *self-injection*. In some simulations a feedback of the electric field of the electrons onto the electric field of the bubble has been observed. This leads to a distortion and elongation of the bubble and is referred to as *beam loading*. The bubble regime

3.3. BUBBLE REGIME

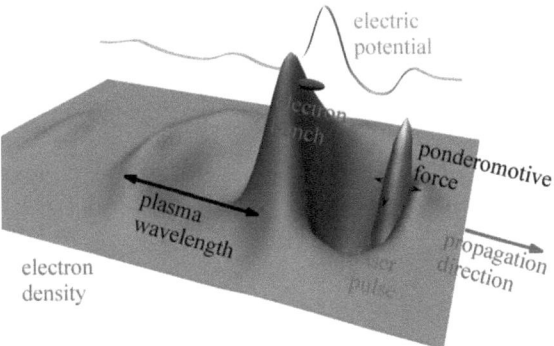

Figure 3.2: Schematic of a bubble acceleration. The laser pulse creates the bubble due to its ponderomotive force. An electron bunch is trapped and accelerated in the electric field of the bubble.

is reached most reliably when the laser pulse is shorter than half the plasma wavelength λ_p (See figure (3.2) on page 31). The electric field has to be higher than about the threshold given by equation (3.23), which usually is given whenever $a_0 \gg 1$. The majority of the statements of chapter (3.2.2) also apply for the bubble regime.

Inside the bubble the electrons are subject to strong electric fields. The confinement of the electron bunch in transversal direction leads to a strongly collimated electron beam with a small angular spread. Also, the transversal fields lead to transversal oscillations of the electrons. However, more interesting in regard to the acceleration are the electric field pointing in forward direction. For an electron density of $10^{18} cm^{-3}$ this field can be of the order of $E_x \approx 100 GV/m$ and therefore is three orders of magnitudes higher than it would even be possible in a conventional accelerator.

The accelerating field is almost uniform in transversal direction. Along the longitudinal axis though it has the highest values close to the back of the bubble and becomes weaker towards the middle of it. This volume of space corresponds to the accelerating phase of this single plasma oscillation. Close to the laser pulse the sign of the electric field changes defining the decelerating phase. If electrons gain enough energy to become faster than the laser pulse and if the laser pulse is not depleted early, the electrons can reach this decelerating phase of the electric field. This leads to dephasing as described in chapter (3.2.3). This electric gradient limits the maximum energy gained by the electrons (compare with equation (3.28)), but also matches the velocities of the electrons. Slow electrons in the back of the bubble are accelerated stronger than faster ones which already have travelled closer towards the center of the bubble where the accelerating fields are weaker. This not only leads to strong acceleration on short length scales, but also to strong collimated beams with small velocity spreads or quasi mono-energetic electrons.

As shown by Kostyukov et al. in [2004, Jun] the accelerating field E_x is almost linear as a function of the relative coordinate $\zeta = ct - x$ and constant as a function of the transversal distance r from the axis. The maximum of E_x is reached for $\zeta = R_B$ where R_B is the bubble radius, assuming a spherical bubble. The radial field E_r and the azimuthal magnetic field B_θ have a linear behaviour with transversal radius r. The values of the fields are

$$E_x \approx \frac{k_p \zeta}{2} E_0 \qquad (3.32)$$

$$E_r \approx \frac{k_p r}{4} E_0 \qquad (3.33)$$

$$B_\theta \approx -\frac{k_p r}{4} E_0. \qquad (3.34)$$

The linear radial fields lead to betatron oscillation of electrons around the propagation axis. During these oscillations electrons emit synchrotron radiation. This will be discussed further in chapter (3.3.2). As Kostyukov et al. found in [2004, Jun] the condition for spot size and laser amplitude in order to reach the bubble regime is

$$k_p R \approx \sqrt{a_0}, \qquad (3.35)$$

while the laser duration τ has to follow

$$\tau \leq \frac{R}{c}. \qquad (3.36)$$

While equation (3.36) is valid for the bubble regime, which usually deals with short laser pulses, equation (3.35) also applies for pulses larger than the plasma wavelength λ_p. This however leads to a regime of self modulation of the laser pulse which will not be subject of this work.

A theoretical model together with extensive simulations was published by Pukhov and Meyer-ter-Vehn in [2002, Mar] which led to the development of scaling laws for the bubble regime by Gordienko, Pukhov et al. in [2005]. These publications are the basis for my work presented in chapter (5.1). It should be pointed out, even though it will not be part of this work, that another theoretical theory was developed by Lu et al. in [2006, Apr] leading to slightly different results concerning scaling laws for the electron acceleration than the ones given by Gordienko, Pukhov and Meyer-ter-Vehn.

3.3. BUBBLE REGIME

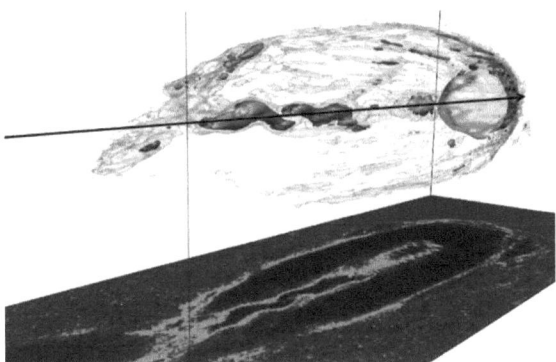

Figure 3.3: Bubble outlines (blue) created by laser pulse (yellow) moving along propagation axis (black). High electron densities are denoted in red. The trapped electrons oscillate around the propagation axis in the plane given by the laser polarisation.

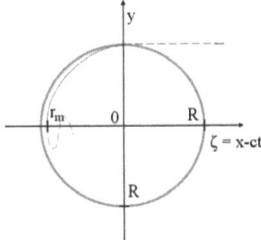

Figure 3.4: Scheme of electron trapping inside the bubble (green). Electrons start in rest at $y = R$, $\zeta = 0$ and are accelerated into the bubble until they reach the turning point r_m.

3.3.2 Electron movement inside the bubble

The trapping of electrons inside the bubble has been studied by Kostyukov et al. in [2009, Jan]. This study included an analytical analysis based on numerical simulations in order to understand where electrons around the bubble start their movement and what condition has to be matched by them in order to get trapped. Simulations suggest that the trajectory of electrons which get trapped start at the side of the bubble at $y = R$, $\zeta = 0$ with a velocity of $\vec{p} = 0$ (compare figure (3.4) on page 33). They are then accelerated towards the propagation axis until they reach the axis at $\zeta = r_m$. At this point their forward velocity has to be higher than the wake field phase velocity. For the spherical bubble it was shown that the condition for trapping is

$$\frac{\gamma_b}{R} \lesssim \frac{1}{\sqrt{2}}, \tag{3.37}$$

where γ_b is the Lorentz-factor of the back of the bubble.

Inside the bubble the trapped electrons follow the electric fields not only in forward direction but also in transversal direction. The transversal electric field is directed towards the propagation axis and therefore keeps the electrons trapped inside the bubble. In transversal direction the field is uniform and its strength it proportional to the radius of the bubble

$$E_r = \frac{m_e^2 k_p^2 r}{2e}, \tag{3.38}$$

with transversal bubble radius r as shown in [2002, May]. While passing $\zeta = r_m$ the electrons about to get trapped already gained a transversal momentum. Therefore the trapped electrons perform a betatron oscillation during their acceleration. In figure (3.3) on page 33 the oscillation of electrons around the propagation axis is shown as an example. In a small bubble $R \leq \lambda_p$ the strong fields which rise linearly in transversal direction conserve a low transversal momentum spread. This leads to a low emittance of the accelerated electron bunch. The betatron wavelength for this oscillation follows from equation (2.32) on page 16 as

$$\lambda_\beta = \lambda_p \sqrt{2\gamma}, \tag{3.39}$$

with the plasma wavelength λ_p and the gamma factor of the electrons γ. Since the electrons achieve relativistic velocities synchrotron radiation is emitted from the bubble. The cut-off frequency for the radiation can be calculated as

$$\hbar \omega_c = \frac{3}{2} \gamma^3 \hbar c r_\beta \left(\frac{2\pi}{\lambda_\beta} \right)^2 \tag{3.40}$$

with r_β being the oscillation amplitude as been shown by Kostyukhov et al. in [2003(Ko)].

3.3.3 Scaling laws for bubble acceleration

With analytic discussion and the use of similarity theory (see the book by Connor and Taylor from [1977] and the paper by Kadomtsev from [1975]) scaling laws for the energy gain of electrons, the number of electrons trapped inside a bubble and dependencies of laser parameters have been found by Pukhov and Gordienko in [2006, Oct]. Because these laws were derived by using similarity theory, which includes dimensional analysis, for some variables only parametric dependencies could be calculated. Dimensionless factors could not be determined analytically. Most of these undetermined variables already have been approximated with simulations. One of the main statements given by Pukhov and Gordienko is that similar states of laser-plasma-interactions are characterised by a corresponding *similarity parameter*

$$S = \frac{n_e}{a_0 n_{crit}}. \tag{3.41}$$

In the regime of relativistically underdense plasma $S \ll 1$ a bubble can be formed which traps

$$N_{mono} \approx \frac{1.8}{k_0 r_e} \sqrt{\frac{P}{P_{rel}}} \tag{3.42}$$

electrons and accelerates them to energies of

$$E_{mono} \approx 0.65 m_e c^2 \sqrt{\frac{P}{P_{rel}}} \frac{c\tau}{\lambda}. \tag{3.43}$$

In these equations $r_e = e^2/(m_e c^2)$ is the classical electron radius, $k_0 = 2\pi/\lambda$, P is the pulse power, while $P_{rel} = m_e^2 c^5/e^2$ is the relativistic power unit. The electrons need to be accelerated over a length of

$$L_{acc} \approx 0.7 \frac{c\tau}{\lambda} Z_R \tag{3.44}$$

with the Rayleigh length $Z_R = \pi R^2/\lambda$ in order to gain this energy.

The principle of similarity can be seen, if different simulations with the same similarity parameter. In figure (3.5) on page 36 two simulations with different S-parameter are shown as a reference. Even though, in both cases a bubble is formed by a laser pulse of $a_0 = 32$, the overall shape of the bubbles differs significant from one another. If instead simulations with matching S-parameter are compared, as in figure (3.6) on page 36 with $S = 1.44 \cdot 10^{-3}$, differences between the bubbles are difficult to spot. Even though all simulations are done with laser pulses ranging between $a_0 = 32$ and $a_0 = 512$, the shape and size of the resulting bubbles is more or less the same.

In the equations (3.42), (3.43) and (3.44) one parameter is the focal spot size R. The spot size is given by equation (3.35) on page 32

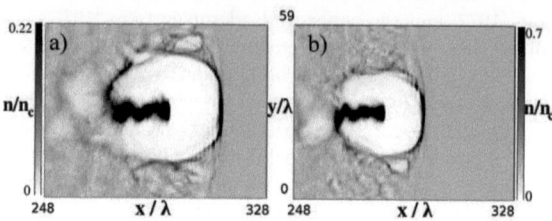

Figure 3.5: *Cut through density profile of two bubble simulation through the propagation axis. Both cases with $a_0 = 32$. S-parameter and electron density correspond to the spot size. a) $R = 5\lambda$ and $S \approx 1.44 \cdot 10^{-3}$, b) $R = 4\lambda$ and $S \approx 2.3 \cdot 10^{-4}$*

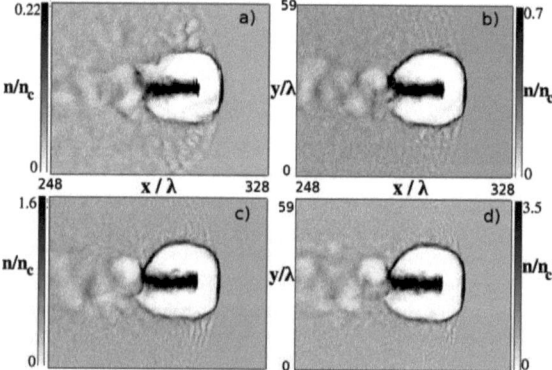

Figure 3.6: *Cut through density profile of bubble simulations through the propagation axis. All simulations were done for $R = 5\lambda$, and similarity parameter $S \approx 1.44 \cdot 10^{-3}$, but different intensities and electron densities. The four cases are a) $a_0 = 32$, b) $a_0 = 128$, c) $a_0 = 256$ and d) $a_0 = 512$.*

3.3. BUBBLE REGIME

$$k_p R \propto \sqrt{a_0},$$

with the plasma wave number $k_p = \omega_p/c$. It also gives an upper limit for the pulse duration τ by equation (3.36) on page 32

$$\tau \leq \frac{R}{c}.$$

In fact Esarey showed in [1996] that the maximum acceleration field is gained when

$$\tau = \frac{\lambda_p}{2c}. \tag{3.45}$$

The pulse power P in equations (3.43) and (3.42) for Gaußian pulses is proportional to R^2 for both linear and circular polarised pulses. For circular polarised pulses one can show that

$$P[GW] \approx 43 \left(\frac{a_0 R^2}{\lambda^2}\right)^2. \tag{3.46}$$

Chapter 4
Particle-In-Cell simulation

4.1 Introduction

In plasma physics, one is often confronted with scenarios where the dynamics of a plasma consisting of many particles are relevant. In laser plasma interaction with densities at about the critical density for a laser the electron density alone can be of 10^{21} particles per cube centimetre. In a volume which at least includes a laser pulse of a size of several micrometre in every direction this would lead to a total of about 10^{13} particles in this volume. This high number of particles leads to computational problems if one would try to simulate such a scenario with each particle modelled individually. Even if one tries to model the plasma as a whole problems arise. As an example A fluid-like model fails to describe situations in which the plasma acts very turbulent. The resulting individual particle motions are important for instance during wave-breaking, which is vital to bubble acceleration (see chapters 3.2.2 and 3.3). One solution to the problem of handling a huge number of particles in electromagnetic fields is the *Particle-in-Cell* (PIC) algorithm. Of course the PIC algorithm can be applied to other fields of physics, as well, but PIC-simulations are well known in particular for their application in plasma physics. In this chapter I would like to explain the concept of the PIC method. For this, first one has to understand how to describe a plasma analytically. Most of the information given here also can be found in the review paper by Verboncoeur from [2004, Oct] and the lecture notes by Pukhov from [1999, May].

From the definition of plasma it is obvious that the plasma can be characterised by the equations (2.3) to (2.6) on page 12. From equation (2.3) and (2.5), with only the curl free part of equation (2.3), it follows the *charge continuity equation*

$$\frac{\partial \rho}{\partial t} + \nabla \cdot \vec{j} = 0, \qquad (4.1)$$

which gives the evolution of the charge density as a function of the current. This means that if equation (4.1) is satisfied at every point in time and if Gauss' Law (equation (2.5)) if fulfilled at any time, then Gauss' Law is automatically fulfilled at every point in time, too. Because of equation (2.4), the same is true for the magnetic field. Thus it is only necessary to find a initial state in which the equations (2.5) and (2.6) are satisfied and from there on we only need to consider the equations (2.3) and (2.4) which give us the time evolution of the system. The current $\vec{j}$ is given by the relativistic Boltzmann-Vlasov equation

$$\frac{\partial f}{\partial t} + \frac{\vec{p}}{m\gamma}\nabla f + q\left(\vec{E} + \frac{\vec{v}}{c} \times \vec{B}\right)\nabla_{\vec{p}}f = C, \qquad (4.2)$$

where $f = f(x, \vec{p}, t)$ is the single particle distribution function of a particle of mass m, charge q, momentum $\vec{p}$ and Lorentz factor γ. C represents the collision term, which usually is part of the Boltzmann-Vlasov equation. However, since in this work relativistic under-dense laser-plasmas are of concern we expect fast electrons, very small cross section for collisions and therefore collision-less plasmas. This means it is save to set $C = 0$. With the equations (2.5) and (2.6) satisfied for the initial condition of a system only the equations (2.3), (2.4) and (4.2) have to be calculated in order to fully describe a system. Unfortunately the Bolzmann-Vlasov equation is a six dimensional partial equation and as such rather complicated to solve. It can be solved numerically using finite differences on an eulerian grid, but this can be very time consuming even if done for just one spatial dimension. An important reason, why these *Vlasov-Codes* are so performance consuming is the fact that the phase space distribution of a plasma does not cover the whole phase space. Which part of the phase space actually is of interest, usually cannot be judged initially. Therefore the eulerian grid covers a vast volume of phase space, which is of no interest for the physical problem. Still a vast part of the phase space is used only to make sure, that the whole phase space distribution of the plasma is covered as well. This is illustrated in figure (4.1) a) on page 41.

A way to approach this problem with less computational costs is to sample the distribution function $f(\vec{x}, \vec{p})$ in a statistic fashion with N *Finite Phase Fluid Elements* (FPFE). These FPFE defined by a shape S^{ph} in phase-space at a certain position $(\vec{x}_n, \vec{p}_n)$ and a weight W^{ph} can be used to define an approximation f_f to the phase space distribution as shown in picture (4.1) b) on page 41. The approximated phase space distribution is given by

$$f(\vec{x}, \vec{p}) \approx f_f(\vec{x}, \vec{p}) = \sum_{n=1}^{N} W_n^{ph} S^{ph}(\vec{x} - \vec{x}_n, \vec{p} - \vec{p}_n). \qquad (4.3)$$

Depending on the shape of the FPFE, equation (4.3) can vastly simplify our problem. For instance, if six-dimensional hyper cubes of side lengths δx_i, $i \in \{x, y, z\}$, are chosen, the shape function is

$$S^{ph}(\vec{x}, \vec{p}) = \begin{cases} 1 & \text{, if } (\vec{x})_i - (\vec{x}_n)_i \leq \delta x_i, \forall i, \text{ and } \vec{p} = \vec{p}_n \\ 0 & \text{, else.} \end{cases} \qquad (4.4)$$

4.1. INTRODUCTION

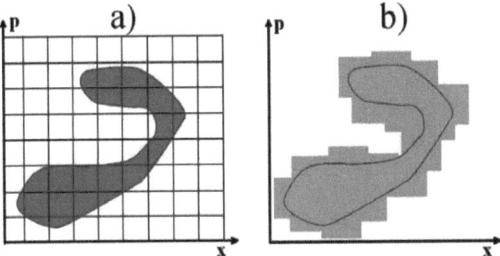

Figure 4.1: *a) Phase space distribution within an Eularian grid. Only the red area is relevant for describing the plasma while the white area is empty. b) The same phase space distribution but now covered with finite fluid elements (orange). The rest of the phase space is ignored.*

Here $(\vec{x})_i$ is the i-th component of $\vec{x}$. Please keep in mind that the 'hyper cube' defined in equation (4.4) in fact is flat in $\vec{p}$−direction, because we only want FPFE with exactly one velocity. These FPFE carry with them the distribution function f_f according to the Vlasov equation (equation (4.2) on page 40) during their movement through phase space given by

$$\frac{d\vec{x}_n}{dt} = \frac{\vec{p}}{m\gamma} \qquad (4.5)$$

$$\frac{d\vec{p}_n}{dt} = q\left(\vec{E} + \frac{\vec{v}}{c} \times \vec{B}\right). \qquad (4.6)$$

The equations (4.5) and (4.6) define the movement of quasi-particles. These particles and their ability to model the development of a given phase space distribution is the key to the PIC algorithm.

The PIC-algorithm is one of the lowest possible level of simplifications and therefore very close to actual physical reality. This is because it solves Maxwell's equations and the Boltzmann-Vlasov-equation for particles very similar to real plasma particles. The only simplification done is the statistical sampling of the phase space distribution. The FPFE correspond to the *macro-particles* in PIC-simulations. These macro-particles, usually defined for electrons and ions separately, have the same mass-to-charge ratio as plasma electrons or ions, but have mass and volume corresponding to a finite cloud of plasma particles. So the macro-particles can be understood as a cloud of ions or electrons co-moving with the same velocity. This seems to be a very intuitive way to reduce the number of particle in a plasma simulation. Furthermore, as it has been shown before, results from such PIC-simulations can be expected to properly model a physical phenomenon, given a 'good' sampling of the phase space distribution. In terms of PIC-simulations, a good sample would be a simulation which sufficiently small and many macro-particles.

This method makes it possible to simulate a dense plasma on modern machines in a very realistic fashion. In PIC simulations the modelling of the plasma is close enough to the real physical situation, so that many important information about the system, like individual particle position, are preserved. The PIC-method has its downsides, too. The most severe one of which is the computational power necessary in order to receive correct results. Even though it has vastly better performance requirements than a Simulation of the dynamic of all plasma particles and even a Vlasov-Code, PIC simulation can become so costly in regard of computational power that for many applications only large computer cluster are able to handle these simulations.

4.2 Algorithm

In this section I would like to explain the PIC algorithm in more detail. For that I will present a basic scheme for a PIC code as it can be found in the literature (for instance the review paper by Verboncoeur from [2004, Oct] or the lecture notes by Pukhov from [1999, May]). Of course a PIC code can be implemented differently, for instance with different solver-methods or adaptive grids instead of Eulerian ones. The method presented here though is a very common one and the basis for even many of the more sophisticated codes in use at the moment.
In a PIC simulation a three-dimensional domain is defined. The domain includes the whole space, which is going to be part of the simulation and everything inside of it. In this domain, particles will be represented continuously in space and velocity while fields are defined on discrete positions. These discrete positions usually are given by an Eulerian grid dividing the simulation domain into three-dimensional cells. In case of the very common Yee-grid, the electric fields are defined on the centre of the surfaces of one grid cell while the magnetic fields point along the edges (see figure (4.2) on page 43 for a explanation of one grid cell). Here the first advantage in regard of performance over Vlasov-Codes becomes clear, since the grid in PIC-simulation has only three dimensions instead of six. Starting from initial conditions for particle positions and velocities, fields as well as particle positions and velocities are advanced sequentially to discrete points in time. If we assume these points in time to be equidistant with difference Δt between two adjacent points t_i and t_{i+1} then usually particle velocities and positions have a time difference of $\Delta t/2$. This way the *leap frog* scheme, a second-order accurate centre difference scheme, can be implemented in order to integrate the equations of motion. The equations for particle positions and velocities in finite form for the leap frog method are

$$\frac{\vec{v}^{t+\Delta t/2} - \vec{v}^{t-\Delta t/2}}{\Delta t} = \frac{q}{m}\left(\vec{E}^t + \frac{\vec{v}^{t+\Delta t/2} + \vec{v}^{t-\Delta t/2}}{2\gamma^t} \times \vec{B}^t\right), \tag{4.7}$$

4.2. ALGORITHM

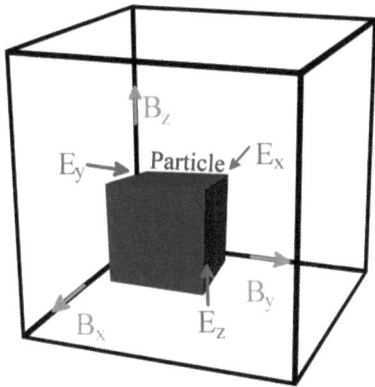

Figure 4.2: *General structure of a Yee-grid cell. Electric fields components (red) are calculated on the surfaces of the cell, magnetic field components (green) on the edges. Particles are placed continuously inside the grid.*

$$\frac{\vec{x}^{t+\Delta t} - \vec{x}^t}{\Delta t} = \frac{\vec{u}^{t+\Delta t/2}}{\gamma^{t+\Delta t/2}}. \tag{4.8}$$

The leap frog method is illustrated in figure (4.3) on page 44. There the circles on the time line correspond to times, at which particle velocities, currents and magnetic fields are defined while the squares represent particle positions, densities and electric fields. Given a starting position at time $t = t_0$ all the values for fields and particle velocities and positions at t_0 and $t_0 - \Delta t/2$ are known (black box). With them equation (4.7) gives the velocities at $t_0 + \Delta/2$, which than can be used in order to calculate the currents and the magnetic fields. Knowing all the values from $t = t_0$, $t_0 - \Delta t/2$ and $t_0 + \Delta t/2$ (orange box in addition to black box) equation (4.8) gives the new positions at $t = t_0 + \Delta t$ (turquoise box). From there one can iterate the process.

The particles in a PIC simulation are macro particles. Instead of simulating every single ion and electron, each macro particle represents many particles of one sort. Since the Lorentz force only depends on the charge to mass ratio, it does not matter how many particles of the same sort are put together into a macro particle. However in order to generate a valid statistic of particle trajectories and an accurate

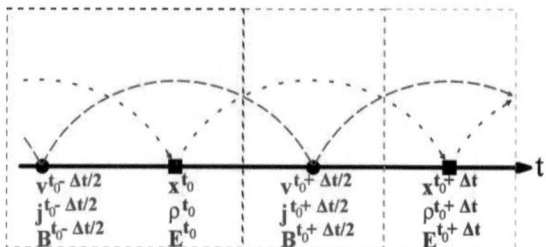

Figure 4.3: *Illustration of the leap frog method. The black box holds the known, initial values for fields and particle positions and velocities. These are used to advance velocities, currents and the magnetic field (orange box) using equation (4.7) on page 42. With this information, the new values for position, particle densities and the electric field can be calculated (turquoise box) using equation (4.8).*

sample of the phase space distribution (see chapter 4.1) the size of the macro particles and therefore the number of real particles corresponding to a macro particle has to be chosen carefully. The simulations presented here are collision free, although particle-particle interactions via electro-magnetic fields of course occur.

Since the focus of this work is electron acceleration within the bubble regime, my goal was to simulate a laser pulse creating a bubble and travelling through a plasma over a certain acceleration length. The total acceleration length in most simulations was much longer than the resolution of the grid, which was in the order of the laser wave length. A simulation domain of the size of the whole acceleration length and with suitable resolution usually would include a large number of cells, difficult to handle with most computer clusters. Therefore I used the moving window technique. With the moving window technique only a small fraction of the whole simulation domain is simulated at a given time, called the simulation box. This box should include every structure one might be interested in. In case of bubble acceleration this mainly includes, the laser pulse, the bubble and the immediate surrounding plasma. The box was co-moving with both pulse and bubble travelling with speed of light. With the reduction of the volume, that is simulated at every time step, the resources needed to complete simulations were reduced drastically.

4.3 Numerical problems and dispersion free solver

One of the major differences of the VLPL to the basic PIC-algorithm is the use of a non-standard field solver. The standard approach for calculating the electric

4.4. PARALLEL COMPUTING

and magnetic field is to use a centred finite difference based on the Yee-grid. Such an approach leads to the *Yee-solver*, which is quiet common in PIC-simulations. However, it is well known that in the Yee-scheme waves travel slower than expected in under-dense plasmas. This effect can for instance reduce the speed of light and therefore electro-magnetic waves significantly.

The *Numerical Dispersion Free Solver* or NDF-solver as presented by Pukhov in [1999] is a field solver which is much more stable than a common Yee-solver. A Yee-grid is a common way to divide the simulation box and to manage the positions at which the fields are represented (see figure 4.2). The details about such a grid and the field solver which usually is used on it are given by Birdsal and Langdon in [1991]. For the Yee-solver in two dimensions one can calculate a numerical dispersion relation as

$$\frac{1}{c^2\tau^2}\sin^2\left(\frac{\omega_k\tau}{2}\right) = \frac{1}{\Delta x^2}\sin^2\left(\frac{k_x\Delta x}{2}\right) + \frac{1}{\Delta y^2}\sin^2\left(\frac{k_y\Delta y}{2}\right). \quad (4.9)$$

This dispersion relation leads to non-imaginary values for the frequency whenever the Courant-Levy criterion is fulfilled,

$$\frac{1}{c^2\tau^2} \geq \frac{1}{\Delta x^2} + \frac{1}{\Delta y^2}. \quad (4.10)$$

A derivation of the criterion and the behaviour of physical (real values for the frequency) and un-physical (imaginary values for the frequency) systems were presented in the review paper by Verboncoeur in [2004, Oct].

The NDF-solver has a different dispersion relation, given by

$$\frac{1}{c^2\tau^2}\sin\left(\frac{\omega\tau}{2}\right) = \frac{1}{\Delta x^2}\sin^2\left(\frac{k_x\Delta x}{2}\right)(b_y + 2a_y\cos(k_y\Delta y))$$
$$+ \frac{1}{\Delta x^2}\sin^2\left(\frac{k_y\Delta y}{2}\right)(b_x + 2a_x\cos(k_x\Delta x)) \quad (4.11)$$

With this dispersion relation one obtains the much better stability condition

$$c\tau \leq \Delta x, \quad (4.12)$$

which only depends on the spatial step in the propagation direction of a wave. With this less restricting stability criterion much larger time steps can be chosen to strongly reduce simulation time.

4.4 Parallel computing

Even on modern computers PIC simulations of extensive systems are more than challenging. As an example I want to estimate the resources needed to simulate the

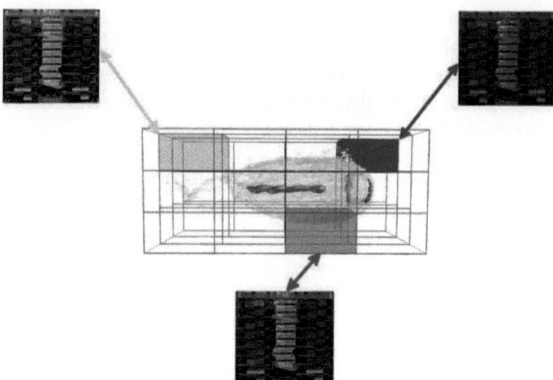

Figure 4.4: *The simulation box is divided into several domains (exemplary in yellow, blue and red). Each cluster node computes the data of exactly one domain.*

immediate surrounding of a laser pulse of wavelength λ travelling through a plasma in 3D. Let us assume a resolution of 0.1λ in direction of pulse propagation and only 0.5λ in transverse direction. This means, that every cell of our PIC-simulation will be a cuboid of a size of $0.03\lambda \cdot 0.25\lambda \cdot 0.25\lambda$. A typical pulse in a bubble acceleration experiment would have a full-width-at-half-maximum of about 15λ transverse and 10λ longitudinal. A good cut-off distance after which the field of such a pulse almost vanishes would be about twice its FWHM in every direction. This would mean, the simulation box would be at about 500 cells in longitudinal direction and 160 in transversal direction. Therefore there are $12.8 \cdot 10^6$ cells in the simulation box. Every cell includes values for the electric field, the magnetic field, the plasma density, the current and the position and velocities of particles inside the cell. If we assume just 8 electrons per cell and no ions, this means that on average, we need at least $3 + 3 + 1 + 3 + 8 \cdot (2 \cdot 3) = 58$ values per cell. With 16 byte per every value, this sums up to $\sim 11 \cdot 10^9$bytes or about 11 Gigabytes. This conservative estimate for just the laser pulse and its immediate surrounding helps to understand, that the demands on the RAM of a computer alone are rather high. Given the fact, that usually one is interested in structures surrounding the laser pulse or travelling behind it and that the processor performance cannot be neglected, it is easy to understand, why someone might want to use more than one computer for a PIC simulation.

The basic idea of parallel computing is to divide the problem on several machines. In case of PIC simulations, the whole simulation box would be divided in sub-boxes, which then again would be handled by different machines or nodes of the same machine (see figure (4.4) on page 46).

Each node computes the steps of the PIC-scheme for its domain and then ex-

4.5. VIRTUAL-LASER-PLASMA-LABORATORY CODE

changes informations with its direct neighbours. In figure (4.5 on page 48) two domains, which are part of the complete simulation box, are shown. Both domains overlap with their neighbours. All the field and particle information from this overlap is exchanged with the corresponding neighbour after every simulation step. A intuitive approach in a PIC-simulation is to use cells on every machine dedicated to its overlap region. This way, after every step, all of the contents of the overlap cells is send to the corresponding neighbour domain without the need of checking whether or not a specific particle or field information has left a domain and now belongs to another one.

As mentioned before, parallel computing also is able to provide better processor performance. Since the computational power of an processor is technically limited, a way to more computational power would be to use more than one processor. *Amdahl's law* gives an estimate for the speed-up, the increase of speed, a code might need to perform a certain task. An important factor for parallel computing is the *parallel efficiency*. According to Amdahl's law the speed-up from using more machines for parallel computing strictly depends on the parallel efficiency. This efficiency refers to the percentage of the processes of a code, which can be parallelised. The remaining operations are strictly serial (for instance any process, which needs global informations). Amdahl's law states, that the maximum speed up $S(n)$ for working on n threads behaves as

$$S(n) = \frac{1}{B + \frac{1}{n}(1-B)}, \qquad (4.13)$$

where B is the percentage of strictly serial processes of the code. The term 'threads' refers to the number of parallel instances of one computation, since in fact most computer clusters use multi-threading in order to have more than one domain on every single node. Equation (4.13) points out, that even for an infinite number of threads, there is an upper limit for the speed-up for any non-vanishing value of B. Therefore it seems reasonable to optimise one's code in order to reduce B as much as possible (which is equivalent to increase the parallel efficiency). In fact in case of the code, I used for this work and which is described in chapter (4.5), tests showed in [1999] that the parallel efficiency is more than 90%. This means that with about 256 threads, which I used for most of the simulations presented here, the speed-up is close to the maximum one, given by $S(\infty)$.

4.5 Virtual-Laser-Plasma-Laboratory code

All the simulations presented here have been done with the *Virtual-Laser-Plasma-Laboratory* Code (VLPL) written by A. Pukhov. Details about this codes can be found in his publication from [1999]. The VLPL is a three dimensional Particle-in-Cell-Code written in C++. It was designed for parallel computing using MPI-routines. It is highly object oriented and can easily be modified to simulate different

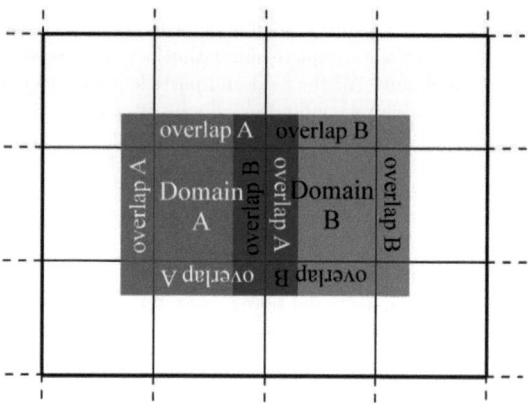

Figure 4.5: *Two domains A and B inside a grid of simulation domains. The domains overlap with the neighbouring ones.*

kinds of particles, shapes of laser pulses, targets and plasmas. For different applications different versions of the code exist, such as one-dimensional or two-dimensional versions as well as a hybrid code between hydro-dynamical- and PIC-simulation, the *H-VLPL*. As mentioned in chapter 4.3 the NDF-scheme is used in order to solve the field equations on a Yee-grid. The macro-particles are rectangular in shape and collision free.

The code also includes radiation reaction. Both classical radiation reaction (corresponding to synchrotron radiation) and quantum-electro-dynamic radiation reaction (corresponding to the creation of gamma quantums and electron positron pairs) can be utilised.

In the VLPL-simulation lengths are normalised by the wavelength λ of the laser pulse (even if a simulation does not include one). Times are normalised by the speed of light and the wavelength c/λ, while both electric as well as magnetic fields are given in numbers of $E_0 = cm_e\omega_L/e$. Field intensity is calculated via the radiant flux

$$\frac{I}{I_0} = \frac{1}{2}(\vec{E}^2 + \vec{B}^2), \tag{4.14}$$

with $I_0 = E_0 \cdot E_0$. The radiant flux I multiplied by $c/2$ gives the intensity in the cgs-gauge.

Chapter 5
Simulations on bubble acceleration

The main part of my work dealt with simulation of electron acceleration within the bubble regime. Here I would like to present the main results of this work. All the simulations preseted here have been done with the *VLPL*-Code using the *NDF*-solver for calculating the field (see chapter 4). Both VLPL-Code and the NDF-solver were presented by Pukhov in [1999]. In the following the natural units of the VLPC-code will be used, which means that all length scales are given in order of the Laser wave length λ, time is given in λ/c, energies in MeV and electric and magnetic fields in $cm_e\omega_{Laser}/e$.

5.1 Comparison of scaling laws

The scaling laws for the bubble regime (chapter 3.3.3) were the central part of my work. My first goal was to investigate whether or not the scaling laws can be validated using PIC-simulation and in which parameter range simulations match theoretical predictions. This included a parameter scan over the possible input values of the scaling laws as well as an investigation of possible reasons for the theory to not match the results from the simulations. Although Before any of this could start, I first had to determine dimensionless factors in the equations (3.42), (3.43) on page 35 and (3.35) on page 32. These factors cannot be derived with the theory used to obtain the equations, but have to be estimated for every laser pulse shape separately.

The equations (3.42) and (3.43) give the values for the number of the accelerated electrons and their energy. However, the pre-factors used in both equations come from simulations using a pulse envelope of

$$a_{orig}(t,r) = a_0 \cos\left(\frac{\pi t}{2\tau}\right) \exp\left(-\frac{r^2}{R^2}\right). \tag{5.1}$$

In this work I used a pulse, which had the envelope

$$a_{new}(t,r) = a_0 \frac{1}{2}\left(1 + \cos\left(\frac{\pi t}{2\tau}\right)\right) \exp\left(-\frac{r^2}{R^2}\right). \tag{5.2}$$

The envelope given by equation (5.1) was defined for $-\tau \leq t \leq \tau$ while equation 5.2 is valid for $-2\tau \leq t \leq 2\tau$. Since the cosine has a local minimum at $t = \pi$, but only an inflexion point at $t = \pi/2$ the pulse given by equation (5.2) drops down smoother to $a = 0$ than the one given by equation (5.1). This changes the pre-factor of both equations (3.42) and (3.43), but features a pulse envelope, which is closer to real laser pulses found in experimental set-ups.

For my simulations I used a grid with different lengths and resolution ranging from 0.1λ to 0.5λ in transversal direction. The exact value was depending on the size of the laser pulse and therefore of the whole simulation box. Even though the resolution of 0.1λ was desirable, with large simulation boxes it was not feasible to use it. As longitudinal resolution I chose 0.05λ since lower resolution tend to lead to numerical dispersion and had a strong impact on the resulting electron energies. The time step was chosen according to the grid resolution and the *NDF*-scheme (equation (4.12) on page 45). In this simulations a laser pulse of the shape given by equation (2.13) travelled through a linear density gradient into a uniform pre-ionised plasma (see figure (7.2) on page 87). The electrons did not have any initial velocity while the ions were considered to be static leading to a uniform electric field in the background. This simplification was done, because laser pulse and electrons involved in the bubble acceleration were both much faster than any ions. Test simulations with ions modelled as PIC-particles did not change the outcome in any significant way and therefore validated this simplification.

At this point a parameter scan would include four free parameters. The pulse amplitude a_0, the pulse width R, its duration τ and the plasma electron density n_e. In order to reduce the number of free parameter, I always chose 'spherical' laser pulses. Equation 3.36 on page 32 gives an upper limit for the pulse duration. In the following I always set

$$\tau = \frac{R}{c}. \tag{5.3}$$

This makes sense in order to maximise the energy gain from a laser pulse of spot size R. Also, in experiments the pulse duration usually is a limiting factor, since it is challenging to create very short pulses. Here it should be kept in mind that the pulse duration still should be so small as to create a laser pulse shorter than half of a plasma wave length. Equation (5.3) reduces the number of free parameters in the equations 3.43 and 3.42 to three, namely the spot size R, the electron density n_e and the laser amplitude a_0. For the parameter scan the set of now three free parameters can be reduced further using (3.35) on page 32. This gives a correlation between R, n_e and a_0 making one of the three a function of the other two. In order to use R as a function of n_e and a_0, the constant in equation (3.35) has to be determined exactly.

5.1. COMPARISON OF SCALING LAWS

Figure 5.1: *Laser pulse with a too small spot size, while travelling through a plasma. The pulse defocusses similar to a pulse in vacuum.*

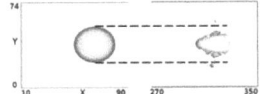

Figure 5.2: *Laser pulse with a too big spot size, while travelling through a plasma. The pulse is quenched by the self-focussing effect.*

5.1.1 Focal spot size

Since equation 3.35 only gives a parametric dependency, the proportional constant has to be determined. The focal radius R is important for the largest possible energy gain for electrons as well as for the wave breaking. This can be seen in the publication by Pukhov in [2005] and in the equations (3.42) and (3.43) on page 35 and equation (3.46) on page 37. A thus optimal chosen radius R also would lead to situation where the relativistic self-focussing effect and defocussing are in balance due to the fitting plasma wavelength (see chapter 2.2.2). This way both laser spot size and the bubble itself keep their shape. For radii not chosen correctly the pulse shape and with it the shape of the bubble change over time, which decreases the energy gain of the electrons. The exact value of R also is important for a comparison of theoretical predictions and numerical simulations as discussed in the introduction of this chapter.

My first approach was to find spot radii which lead to laser pulses travelling through the plasma without changing their transversal shape. In figure 5.1 and 5.2 two cases are presented in which the laser pulses did not match the optimal spot size. Figure 5.3 shows a laser pulse which is stabilised by the self-focussing effect and passes its Rayleigh-length without significant changes in its transversal size. The longitudinal shape however changes. This is easy to understand, since the front of the pulse is travelling through a denser medium than the back. Due to the density and therefore the diffraction index being higher at the front of the pulse, the back is slightly faster. This quenching of the laser pulse however does not hinder the bubble acceleration. On the contrary, a short pulse confined to the very front of the bubble interior does not effect the trapped electrons as much as a more extended one.

Of course, the laser pulse being stable does not mean that its pulse width is exactly the optimal one given by equation (3.35). However, this search gave us a first guess for the optimal radius, which then later (chapter 5.1.2) could be confirmed by comparing acceleration length, number of trapped electrons and energy gain from simulations with the predictions from the theory. For the analysis of the development of the spot size I determined the full-width at halve-maximum (FWHM) of the laser pulse in transversal direction during simulations. Strong variations from

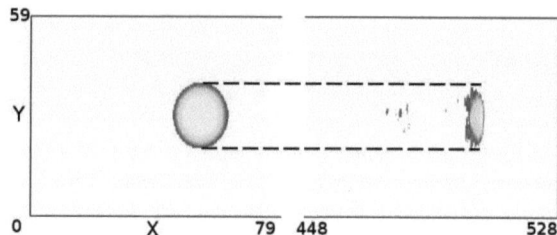

Figure 5.3: *Laser pulse with too small spot size, while travelling through a plasma. The pulse stays stable in transversal direction even long after passing its Rayleigh-length.*

its initial value and strong fluctuations can be considered to be an evidence for a not-optimal initial radius. For this I implemented a routine into the VLPL-code, which tracked the pulse and determined its FWHM in simulation time. This way I avoided saving all the three-dimensional field information at enough time steps in order to make a reasonable analysis.

Both figure (5.4) and (5.5) give three simulations each for three different initial pulse widths. Both sets of simulations have been done for the same similarity parameter $S = 0.001$. The plasma in this simulations was a electron plasma and a positively charged background with a linear density gradient in the beginning and a constant density plateau afterwards. In both cases the pulse with initial width $R = 6\lambda$ seems to be the most stable one while travelling through the density plateau. The reason for the peaks at the beginning of the simulations are the strong reaction of the pulse to the density gradient. The pulse shape, while travelling through the gradient, was strongly distorted from a gaußian shape as can be seen in figure (5.6) on page 53. This distortion of course effects the calculated FWHM, wich was calculated after fitting a Gaußian distribution through the intensity distribution. After passing the gradient, the pulse regained a much more gaußian shape. The peaks at the end of the simulations occur because of depletion. The laser pulse lost almost all its energy, which again leads to a pulse profile which does not match a gaußian shape very well any more.

With this first guess for an optimal spot size and introducing a new constant C_R with the equation (3.35) on page 32

$$k_p R = C_R \sqrt{a_0}, \tag{5.4}$$

we can use the equations (2.12) and (2.21) on page 15 to derive a equation for the

5.1. COMPARISON OF SCALING LAWS

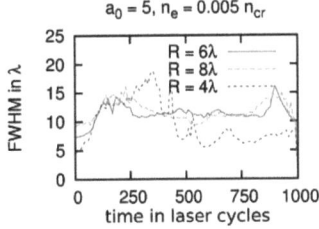
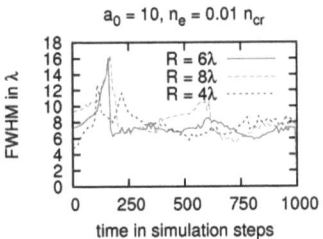

Figure 5.4: *Exemplary set of simulations in order to find a stable laser pulse. $R = 6\lambda$ seems to be the best of the three choices. The peaks at the beginning of the simulation is due to the density gradient, the plasma in the simulation had. The peaks at the end are results of the depletion of the laser pulse.*

Figure 5.5: *Similar to fig. 5.4, but for a different set of a_0 and n_e. The S-parameter in both figures is the same.*

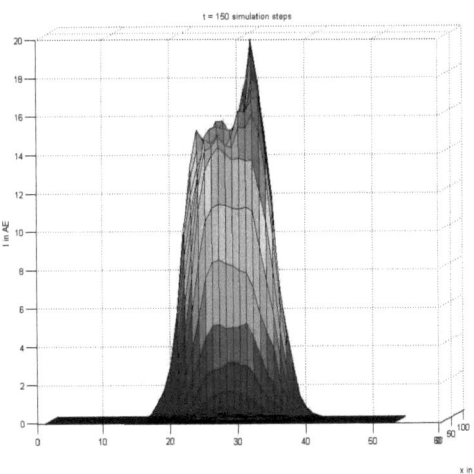

Figure 5.6: *Intensity distribution of a laser pulse while travelling through a density gradient. Strong variations from the initial gaußian profile are visible.*

spot radius as a function of the similarity parameter S. With

$$k_p = \frac{\omega_p}{c} = \sqrt{\frac{4\pi n_e e^2}{m_e c^2}} \tag{5.5}$$

$$\overset{eq.2.21}{\Rightarrow} k_p\sqrt{n_{crit}} = \frac{\sqrt{n_e}\omega}{c} = 2\pi\frac{\sqrt{n_e}}{\lambda} \tag{5.6}$$

this leads to

$$\Rightarrow \frac{R}{\lambda} = C_R \frac{\sqrt{n_{crit}}}{\lambda} \frac{\sqrt{a_0}}{\sqrt{n_{crit}}k_p} = \frac{C_R}{2\pi}\sqrt{\frac{n_c a_0}{n_e}} = \frac{C_R}{2\pi\sqrt{S}}. \tag{5.7}$$

If we now put into equation (5.7) that for a similarity parameter of $S = 0.001$ the optimal spot size would be $R \approx 6\lambda$ we get that $C_R/2\pi \approx 0.19$ or

$$\frac{R}{\lambda} \approx 0.19 \frac{1}{\sqrt{S}}. \tag{5.8}$$

Of course this approximation of C_R only holds for this choice of pulse envelope, but it should be correct regardless of the choice of S. To proof that this approximation to C_R is a good one, I checked whether results for electron energy gain and number of trapped electrons from my simulation fit predictions from the theory (equations (3.42) and (3.43)) using equation (5.8). The rather positive results can be seen in detail in the following chapter.

Another information equation (5.8) holds is the fact that the optimal spot size R and the similarity parameter S can be used interchangeable. Every similarity parameter S corresponds to exactly one optimal spot size R. It is important to distinguish between the optimal spot size R and any other Laser spot size r_L one might choose for an experiment. While the definition of S (equation (3.41) on page 35) allows any value for S for any given spot size r_L by changing the Laser amplitude or the electron density, S defines exactly one optimal spot size R.

5.1.2 Energy gain and trapped particles

Using the equation (5.8) on page 54 and the definition of the similarity parameter (3.41) on page 35 the set of parameters for a parameter scan is reduced to two. For any chosen spot size R equation (5.8) gives the appropriate similarity parameter S. The second parameter, which has to be chosen is either the laser amplitude a_0 or the electron density n_e. The other one is determined by equation (3.41) on page 35. In the following I always used a_0 as a free parameter, calculating n_e from equation (3.41).

With this preparations I was able to compare the energy from electrons from simulations with the predictions from equation (3.43). For this I calculated the

5.1. COMPARISON OF SCALING LAWS

ratio Q_E

$$Q_E = \frac{E_{\text{sim}}}{E_{\text{theory}}} \tag{5.9}$$

of the electron energy from simulations E_{sim} over the energy predicted by the theory E_{theory}. A ratio of $Q_E = 1$ therefore means, that theory and prediction perfectly matched, while higher values of Q_E correspond to simulations during which the electron energies exceeded the predictions. In my simulation a different laser pulse shape was used then in the simulations which lead to the pre-factor in the original equation (3.43). The dependence of the expected energy on the exact shape of a laser pulse can be seen if one compares the figures 5.8 a) and (5.7). Both figures show a parameter scan of Q_E as a function of R and a_0, using the results of 30 simulations each. Figure (5.7) on page 57 shows the ration Q_E using the original equation (3.43) for the energy of the accelerated electrons. As can clearly be seen, even though the Laser envelopes are similar, overall a significantly higher energy is gained with the new shape. Using a least-square fit method in order to match the parametric dependency

$$E_{\text{parametric}} = C_E m_e c^2 \sqrt{\frac{P}{P_{rel}}} \frac{c\tau}{\lambda} = \tilde{C}_E a_0 \frac{R^2}{\lambda^2} [MeV] \tag{5.10}$$

to the energies gained in my simulation presented in figure (5.7), I could calculate the new factor $\tilde{C}_E \approx 1$. This lead to the equation

$$E_{\text{mono, new}} \approx 0.93 m_e c^2 \sqrt{\frac{P}{P_{rel}}} \frac{c\tau}{\lambda}. \tag{5.11}$$

for the energy of the trapped electron bunch, using an envelope of the form given by equation (5.2) on page 49.

The ratio Q_E with the predicted energy given by equation (5.11) can be seen in figure (5.8) a) on page 58. In this plot in contrast to figure (5.7) there is a wide range of parameters at which theory and results from simulations fit almost perfectly. At very small laser spot sizes my simulations achieved slightly less energy then expected in particular for very high laser amplitudes $a_0 \geq 128$. At low amplitudes of $a_0 \leq 4$ I found deviation from the theory. In the plot one can see higher energies at some of the radii and lower energies at $R = 2\lambda$. The reason for this can be seen by comparing two snapshots from different simulations. In figure (5.7) c) the density- and intensity distribution for a simulation at $R = 6\lambda$ and $a_0 = 128$ is shown. There a well defined bubble can be seen with trapped electrons and a broken wake field behind the laser. In figure (5.7) b) in contrast to that, one can see a laser pulse followed by a periodic wake field. The majority of electrons is neither inside of the bubble or in front of the laser pulse, but to the sides of the cavities. Some electrons are trapped inside the first cavity, but their number is quiet low. Still though figure (5.9) on page 59 shows that electrons in fact are accelerated in a quasi mono-energetic fashion. This indicates that the simulations at $a_0 = 4$ are situated at the boundaries of the bubble regime. In fact during simulations with low Laser amplitude $a_0 < 4$ or small spot sizes $R < 2\lambda$ almost never a bubble was formed. Figure (5.8) a) shows that for spot sizes of $R \geq 3\lambda$ and for laser amplitudes of $a_0 > 4$ I received very good agreement of energies from the theory and the simulations.

The number of accelerated particles behaved slightly different with the new laser pulse shape, too. I found that for this shape the equation for the number of electrons should be

$$N_{\text{mono, new}} \approx \frac{3.01}{k_0 r_e} \sqrt{\frac{P}{P_{rel}}}. \tag{5.12}$$

With this new factor the ratio Q_N of number of electrons accelerated in a monoenergetic fashion over the expected value $N_{mono,new}$ can be seen in figure (5.10) on page 59. Again the region of $a_0 = 4$ differs significantly from the theory. However, as discussed above, this region is a border of the bubble regime. In the majority of the parameter space I found agreement between theory and simulation, aside from high intensities at low spot sizes. To understand this discrepancy it is useful to have a look at the relative energy spread $\Delta E/E$. ΔE denotes the width of a peak in the energy spectrum, while E is the position of the top of the peak. Figure (5.11) a) on page 60 illustrates this information. At the position, at which the anomaly occurs, much higher relative energy spreads can be found. A Comparison of two energy histograms from similar simulations, in this case $R = 2$ and similarity parameter $S = 0.009$, can be found in (5.11) b) and c). b) shows a good example for quasi mono-energetic acceleration with a well pronounced peak at the highest energy values in this histogram. c) in contrast to this shows a less pronounced peak and further more electrons with higher energies. Both histograms were done after the same acceleration length, but it seems, that in case of c) the bubble already became instable and the electrons started loosing energy. Since in similar experiments the same acceleration length is expected according to equation (3.44), very small spot sizes seem to be another limit to the scaling laws. The wider energy peaks, which are results of an instable bubble, also explain the slightly lower energies in figure (5.8) a) on page 58 at the same parameters.

The reason why the parameter scan ends at radius $R = 12\lambda$ and $a_0 = 512$ is for one, that I could not find any evidence that at high laser amplitudes the scaling laws would not match the simulations. Also, at an amplitude of $a_0 = 512$ we already are in a regime which is rather unlikely to be reached by experiments in the near future. No simulations with larger spot sizes have been conducted for computational reasons. Since the resolution of the simulation grid has upper limits given by the laser- and plasma wave lengths, the number of cells inside the simulation box becomes increasingly larger for large spot sizes. Also, the acceleration length and therefore the simulation time increases proportional to R^2 (equation (3.44) on page 35). Simulations with spot sizes of $R = 12\lambda$ already took seven to ten days each and the majority of the available RAM. Larger spot sizes would need much more

5.1. COMPARISON OF SCALING LAWS

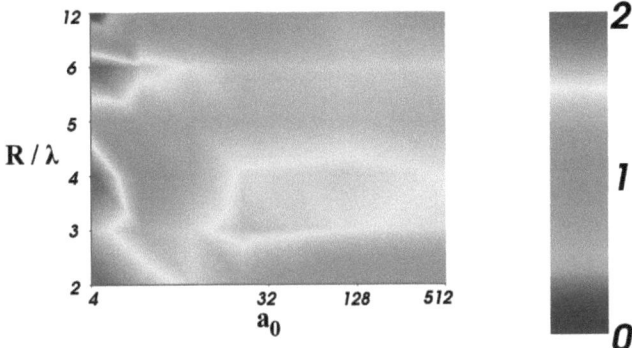

Figure 5.7: *Electron energy ratio $Q_E = E_{sim}/E_{theory}$ using with the original factor according to equation 3.43 on page 35.*

computational power then available or advanced schemes like PIC-hydrodynamic-hybrid-Codes or Lorentz-boosted simulations.

In conclusion I can state that I found lower limits for the parameters a_0 and R at which simulation and theory stop matching and a wide parameter space in which theory and simulation fit well. So far, no upper limits for a_0 and R are known. However, since at very high laser intensities the electrons achieve very high energies, at large values for a_0 synchrotron radiation should be considered.

5.1.3 Radiation reaction

So far my simulations did not include any form of radiation reaction. At low values for a_0 this should not make a difference, but since the parameter scan ranges up to $a_0 = 512$ radiation reaction should be considered. Apparently one of the first points to address is at which point radiation reaction has to be considered in order to make realistic predictions.

At first I only considered classical synchrotron-like radiation reaction without any quantum-electro-dynamic (QED) effects. Radiation reaction in the VLPL-code is a modification of the the particle pusher. At every time step the energy and momentum and their change for every particle is calculated and used to decide whether or not a particle emits radiation. With radiation reaction I obtained the results given

Figure 5.8: a) *Comparison of analytic predictions and results from simulations. The colour represents the quotient of the energy from the simulation over the prediction $Q_E = E_{sim}/E_{theory}$. Green colour was used for a quotient of $Q = 1$ which means, that theory and simulation perfectly match. Red colour and a quotient of $Q > 1$ represents simulations which achieved higher energies than expected from the theory, while blue stands for simulations with lower energy than expected.*

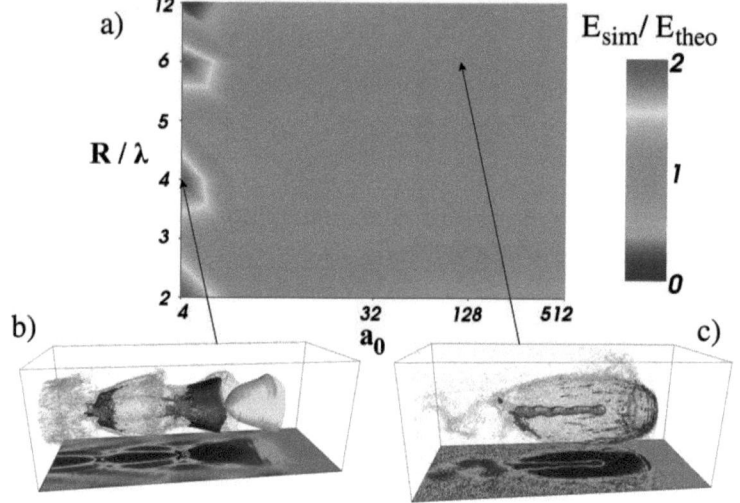

b)*Snapshot from a simulation at $R = 4$, $a_0 = 4$ and $S = 2.25 \cdot 10^{-3}$. The 3D-figure shows the electron density in blue (low density) and red (high density) and the Laser intensity in yellow. as well as a 2D cut of the density through the middle of the box. The picture shows a structure with a periodic, unbroken wave following the laser pulse. Some electrons are trapped inside the first cavity, as can be seen in the 2D-cut, but much more are just following the oscillations.*

c) *Similar to b), but at $R = 6\lambda$ and $a_0 = 128$, which leads to $S = 0.001$. The bubble structure is easy to spot as well as the trapped electrons. The 2D cut beneath the 3D image also shows that the bubble is only followed by some turbulances.*

5.1. COMPARISON OF SCALING LAWS

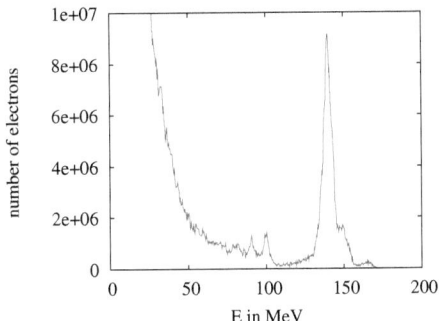

Figure 5.9: *Electron energy corresponding to the situation in figure 5.8 b) on page 58. This histogram shows a well defined peak indicating that electrons are accelerated in a quasi mono-energetic fashion.*

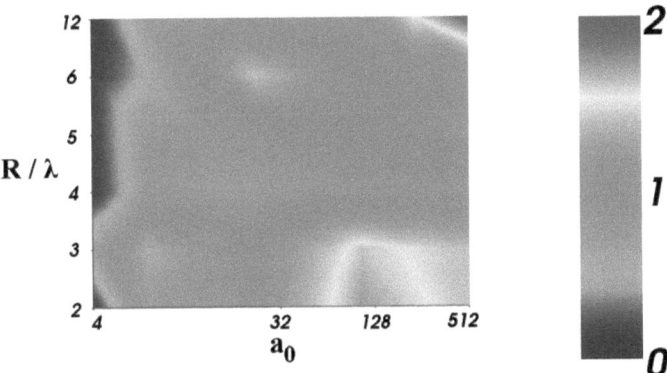

Figure 5.10: *Ratio of the expected number of accelerated electrons over the number of electrons accelerated during the simulations $Q_N = N_{sim}/N_{theory}$.*

Figure 5.11: *a) Relative energy spread of the energy peak of the quasi mono-energetic electron bunch. Green corresponds to a relative energy spread of 10% while red stands for higher spread and blue for less spread.*

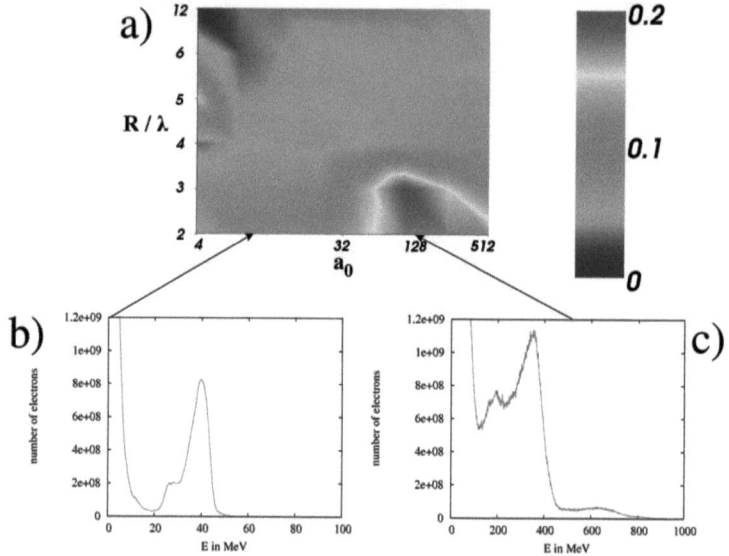

a) *Energy histogram of electrons at $R = 2$, $a_0 = 8$ after a acceleration length of 70λ. A clear pronounced energy peak containing the fastest electrons can be seen.*
b) *Same as a), but at $a_0 = 128$. Here the peak is less pronounced and electrons of higher energy can be seen.*

5.1. COMPARISON OF SCALING LAWS

in figure (5.12). The first guess that radiation reaction does not change the results at low values was validated by this simulations for $a_0 \leq 50$. Above that especially beginning at about $a_0 = 100$ we can see, that the energy gain of the electrons is significantly lower then in the simulations before. Due to synchrotron radiation the electrons inside the bubble loose energy during their betatron oscillation. At very high energies the loss of energy due to radiation reaction also leads to instable bubble formation and propagation. In figure (5.13) on page 62 two sets of histograms are shown. (a) refers to the point ($R = 12\lambda$, $a_0 = 512$) in figure (5.12) (Energy ratio with synchrotron radiation), while (b) refers to (5.8) (the same without synchrotron radiation). As can be seen, the acceleration length, given by the green curve in both cases, is reached at different times, even though both systems are similar. As expected, a) achieved significantly less energy, but also the peak is less well pronounced. In fact, a peak is only visible for a short fraction of the total acceleration length, while without synchrotron radiation, peaked structures are clearly visible for the most part. However, most surprisingly, during the simulation belonging to a), more then ten times as many electrons were accelerated. To help understanding this, figure (5.14) on page 63 shows another cut through the transversal axis of the density profile for both simulations. The simulation without synchrotron radiation (b) shows a well defined bubble with a small stem growing from its back. Contrary to this the simulation with radiation reaction (a) pictures a structure with a well defined front, but a unstable, 'noisy' back. The border between inside and outside of the bubble fluctuates and particles move between both areas. The bubble is strongly elongated and therefore much longer than the plasma wavelength. This seems to be an evidence, that at very high laser intensities, the synchrotron radiation not only limits the energy gain, but inhibits the formation of regular bubble. This behaviour and its resulting trapping of more electrons can be seen in figure (5.15) on page 63. There it becomes evident, that in the area, where lower energies occur than expected (figure (5.12), also much more electrons are trapped.

Synchrotron radiation not only reduces the energy significantly, but it also impairs the energy spread and with this the quality of the electron beam. The relative energy spread with synchrotron radiation can be seen in figure (5.16) on page 64. Starting at about $a_0 = 128$ the energy spread increases far over 10% with increasing a_0.

In case of $R = 12\lambda$ it seems, that the effects of synchrotron radiation occur on much smaller laser amplitudes, comparable to the simulations at $R = 2\lambda$. At this point it is unclear whether or not this behaviour at $R = 12\lambda$ is connected to the worse resolution used in the $R = 12\lambda$ simulations compared to the other ones.

If instead of classical synchrotron radiation the creation of gamma quanta is used the results almost does not differ. The quantum dynamical creation of high energetic photons is implemented in the VLPL-code in a similar manner as the creation of synchrotron radiation. The main difference is the fact, that the gamma-quanta are real PIC-particles, which move and can interact with other particles and

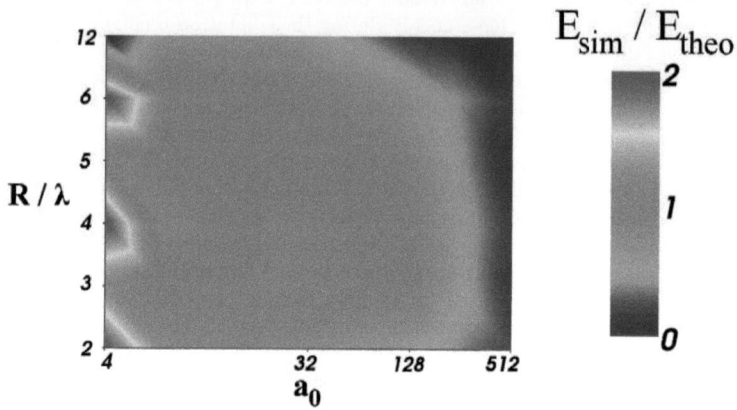

Figure 5.12: *Energy ratio similar to figure 5.8 a) on page 58, but now with classical radiation reaction.*

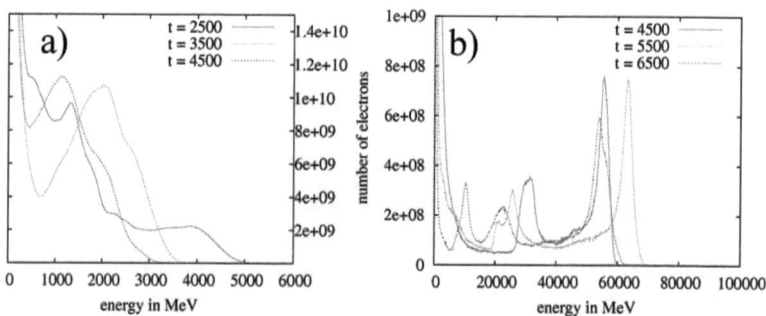

Figure 5.13: *Histogram from two simulations at $R = 12\lambda$ and $a_0 = 512$ with synchrotron radiation (a) and without (b) at different time steps.*

5.1. COMPARISON OF SCALING LAWS

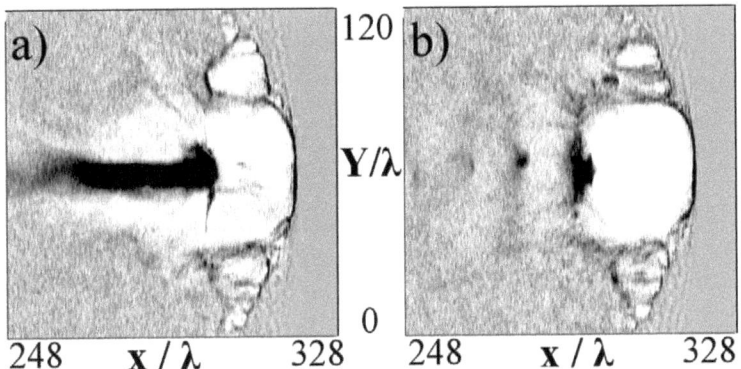

Figure 5.14: *Two cuts through the density distribution of bubble acceleration simulations at $R = 12\lambda$, $a_0 = 512$ and $t = 1000\lambda/c$. One simulation was done with synchrotron radiation (a) and one without (b).*

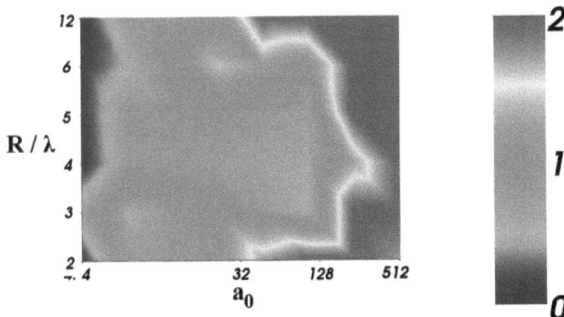

Figure 5.15: *Ratio of the expected number of accelerated electrons over the number of electrons accelerated during the simulations $Q_N = N_{sim}/N_{theory}$. Now with synchrotron radiation.*

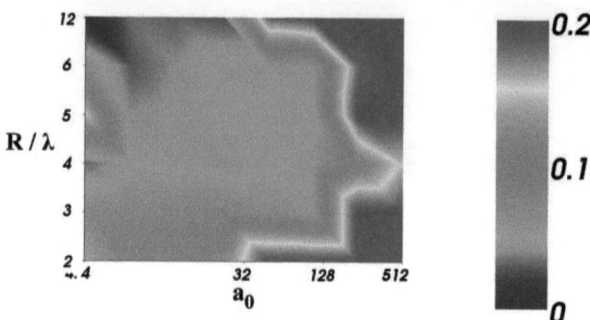

Figure 5.16: *Relative energy spread similar to figure 5.11 a) on page 60, but now with classical radiation reaction. With $R = 12\lambda$ and $a_0 = 512$ no mono-energetic peak could be found in the electron spectrum.*

fields via the generation of electron-positron pairs. Also, whether or not photons are emitted is determined by a Monte-Carlo-like algorithm, based on the energy of the emitting particles. Emitting gamma-quanta reduce the energy of electrons, similar to the effect of synchrotron radiation. One would suspect, that these impact on the electron energy would only occur on very hight electron energies. In figure (5.17) on page 65 it is shown that for the whole parameter space the decrease of the energy of electrons becomes significant about at the same energy as for the classical synchrotron radiation. Furthermore, even the amount of which the electron energies are reduced are comparable for both sets of simulations. From this it seems, that in the regime I studied the difference between classical synchrotron radiation and quantum-electro-dynamical creation of gamma quanta is negligible. An exact study, at which point to use radiation reaction emulating classical synchrotron radiation and at which point QED-effects become more important, therefore seems to be not necessary.

From this an upper limit for the scaling laws seems to be around $a_0 \sim 128$ at which point synchrotron radiation cannot be neglected any more. Below that the energy emitted by the electrons does not significantly influence the acceleration.

5.1.4 Summary

In this section I investigated the scaling laws for the bubble regime (chapter 3.3.3) specifically the energy gain of the electrons and the number of electrons accelerated.

5.1. COMPARISON OF SCALING LAWS

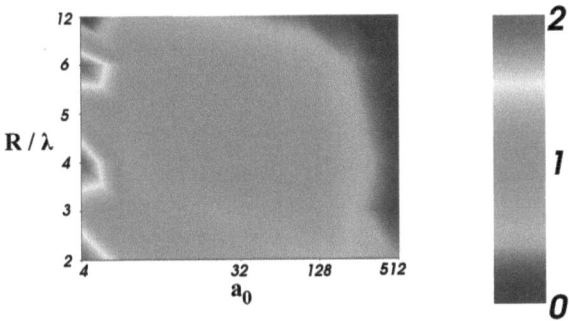

Figure 5.17: *Energy ratio similar to figure 5.8 a) on page 58, but now with gamma quanta creation.*

For this I performed a series of simulation in order to find laser pulse specific, dimensionless factors for the scaling laws. With these factors I conducted a parameter scan in order to find a regime in which theory matches prediction and therefore validate the scaling laws. Having found such a regime I tried to find the limits of this regime and to understand them. As lower boundaries I found that for low laser amplitudes $a_0 \leq 4$ and small spot sizes $R \leq 2\lambda$ the bubble formation became more and more incomplete. The laser pulse instead was followed by a periodical wakefield, which denotes the beginning of a different regime of particle acceleration. An upper limit at $a_0 \sim 128$ is a result of energy loss of the electrons due to synchrotron radiation. Even higher laser amplitudes in certain cases lead to incomplete formation of the bubble. At lower laser amplitudes the radiation emitted by the electrons while betatron oscillating inside the bubble does not influence their energy gain significantly. An upper limit for the spot radius is not known so far. Synchrotron radiation also reduces the quality of the electron beam created by the bubble acceleration by increasing the energy spread of the quasi mono-energetic electron bunch. An additional term, approximating the energy loss at high laser amplitudes due to synchrotron radiation was presented and briefly discussed.

5.2 Comparison with bubble acceleration experiment

The simulations presented in this work not only matched theoretical predictions (see chapter 5.1), but also achieved results which could directly be compared to experiments done in the field of bubble acceleration. One important aspect of laser-plasma simulations is the development of the laser pulse during the simulation. Interpolation of fields, particle movements and positions, the resolution of the lattice and boundary conditions in a PIC-simulation can lead to non-physical behaviour like numerical dispersion, numerical heating, other conservation issues and more. Some numerical issues can be spotted fairly easily while others have a more subtle impact on simulation results. Detailed information about numerical issues with simulations can be found in the book by Birdsal and Langdon from [1985]. Especially the more subtle numerical problems as well as any assumption or simplification a numerical simulation might include make it vital to compare simulation results with experimental data. In this chapter I would like to present some simulations made to reproduce experimental data using the same input parameter as been used by the respective experimentalists.

5.2.1 Pulse development

Since the driver for the wake field, which is the source for the electron acceleration, is a laser pulse I would like to start with a study of the development of a laser pulse travelling through a plasma. This study was done in collaboration with Ariane Pipahl, who published the results in [2010]. All the experimental data presented in this chapter have been obtained by her working at the Arcturus Laser Facility of the Heinrich-Heine-Universität. The initial electric field and its corresponding envelope can be seen in figure (5.18) on page 67. The envelopes have been created by extracting the electric field data from the simulation and Fourier-transforming them along the propagation direction. In figure (5.19) on page 67 three laser envelopes are shown. The initial envelope of the simulated laser pulse after initialisation (grey), the laser pulse after travelling through a plasma of density $n_e = 5 \cdot 10^{18} cm^{-3}$ and the envelope of a laser pulse according to experimental data after travelling through a plasma of density $n_e = 6.6 \cdot 10^{18} cm^{-3}$. In both cases a laser pulse with an amplitude of $a_0 = 4.7$ a duration of $28fs$ and a radius of about $5\mu m$. Both simulation and experiment point out a red shift of the laser pulse with an additional peak at about $925nm$ in the simulation data and at about $900nm$ in the experimental data. This red shift is a result of a self-modulation of the laser pulse. Also, due to the increase of electron density and relativistic electron mass in front of the laser (see chapter (2.2.2)) a change in the different diffraction is created. This increase of density of mass induced by the laser pulse in return leads to a modulation of the electric field. The increase of density in front of the laser pulse can be seen in figure (5.20) b) on

5.2. COMPARISON WITH BUBBLE ACCELERATION EXPERIMENT

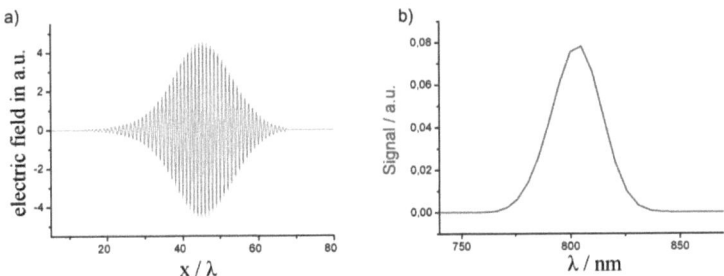

Figure 5.18: *Electric field of a just initialised laser pulse in propagation direction (black) and the envelope as a function of the laser wave length λ (red).*

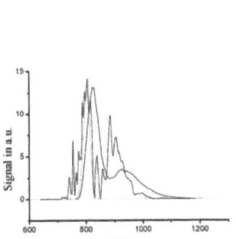

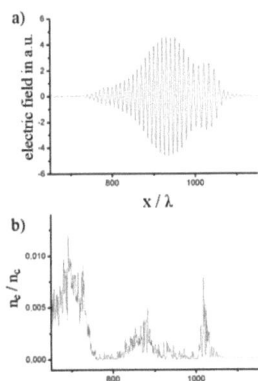

Figure 5.19: *Laser envelope simulated of a laser pulse before (gray) and after (red) travelling through a target of $n_e = 5 \cdot 10^{18} cm^{-3}$. Blue denotes experimental results by Ariane Pipahl in the same regime.*

Figure 5.20: *a) Electric field along the propagation axis. A clear modulation at the front of the pulse can be seen. b) Electron density along the propagation axis. At the same position as the field modulation, an increase of electron density was found.*

page 67. The resulting modulation is found in figure (5.20) a). This modulation is the reason for the red-shift in the envelope. The experimental data show an additional blue-shift, which is a results of ionisation. The electron density in the experiment is a function of time as well as position and the rear part of the laser pulse experienced a much different electron density as the front part. Since during the simulations the gas was fully ionised and therefore ionisation effects did not take place, this behaviour cannot be seen in the simulation data. The overall results seems comparable. Both data sets show a red-shift and a second signal peak at around $900 - 925nm$. The red shift in the simulation though is a more collective one then in the experiment and the position of the second peak and its hight relative to the hight of the first peak differ slightly between simulation and experiment. It should be noted that the positions of the peaks differ only by less than 10% and that the rather collective and smooth shift of the simulation curve to higher wave length most likely is based on the fact that the simulation did not include ionisation. Therefore the simulation was apt to reproduce the physical behaviour of the development of the laser pulse with only slight differences in the red-shift, the exact position and relative hight of the second peak.

5.2.2 Electron acceleration

In chapter 4.1 I stated that PIC-simulation are considered to be very realistic. To check how close to reality the simulations, I conducted, were, I compared my results to data obtained from experiments with corresponding parameters. While comparing electron spectra obtained from experiments with electron spectra from simulations, one should keep in mind that experimental results tend to differ slightly from experiment to experiment. I tried in this chapter to regard a large number of experimental results in order to find typical structures, which I then compared the data from simulations. Also, I would like to point out, that both facilities from which the experimental data originate did not anticipate electron energies of more than $200 MeV$. Both facilities are not equipped to deal with such electrons safely, therefore no data about such electrons were available.

An experimental set-up for a bubble acceleration includes all the equipment for creating a laser pulse, guiding and focussing it on a target of course. The part of the set-up, which was most important for my simulations though, was the target and its immediate surrounding. Typically this would be a ultrasonic gas stream ejected from a gas nozzle inside an evacuated container, the gas chamber. This can be seen schematically and simplified in figure (5.21) on page 69. Since many laser systems are able to emit several pulses within short periods of time, it is quiet useful to have a target, which is replaced quickly between several shots. Furthermore, since the target is going to be ionised (by the laser pulse itself or a 'pre-pulse') and one usually wants to use under-dense targets, a gas stream seems to be a very good choice. This method has a few minor downsides however. The gas stream moves at a ultra-sonic speed, which means that it is almost static in the frame of the laser pulse, but the

5.2. COMPARISON WITH BUBBLE ACCELERATION EXPERIMENT

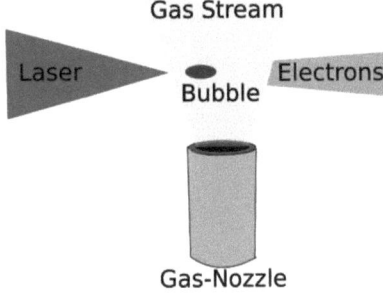
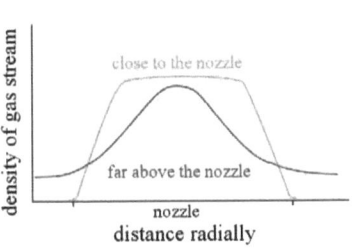

Figure 5.21: *Simplified experimental set-up for a bubble experiment. A laser pulse is focussed on a gas stream on one side. Within the stream, emitted by a gas nozzle, a bubble is formed. After the acceleration, electrons will leave the gas stream on the opposing side.*

Figure 5.22: *Density distribution of gas jet above the gas nozzle. Turquoise denotes the distribution just above the nozzle, while red corresponds to the distribution far above the nozzle.*

target area is replenished between shots. The only downside is, that most gas chambers cannot be evacuated in the same speed the gas is filled in. That means that during later shots in the same experiment the laser interacts with gas even before it is focussed on its target. Also, the density profile poses challenges. The density distribution of the gas stream almost always will have a gradient, which is difficult to control, and not an entirely flat plateau. The profile will become worse with increasing distance to the nozzle as is illustrated in figure (5.22). Unfortunately it is unpractical to focus the laser pulse too close to the nozzle, because the laser pulse can significantly damage the nozzle. In the simulations presented here I modelled the gas stream as a fully ionised plasma with a linear gradient and a flat plateau.

The first experiment was done by Ariane Pipahl at the Arcturus Laser Facility at the Heinrich-Heine-University Düsseldorf. In her experiments she used a laser pulse with a wavelength of $\lambda = 800nm$ a pulse duration $\tau = 28fs = 10.5\lambda/c$, a spot radius of $r_L = 5\mu m = 6.25\lambda$ and an intensity of $I_0 = 4 \cdot 10^{19} W/cm^2$ corresponding to an amplitude of $a_0 = 4.3$. The electron density in this experiment was $n_e = 1.3 \cdot 10^{19} cm^{-3} = 7.44 \cdot 10^{-3} n_c$. The result of one experiment is shown in figure (5.23) on page 70 after an acceleration length of about $2500\mu m$. A distinguished multi-peak spectrum can be seen to energies up to $120MeV$. Electrons of higher energy exist though they do not seem to be mono-energetic. The measurement stopped at $200MeV$, because only electrons with energies of $10 - 200MeV$ were expected and no measurement of higher energies have been conducted. Simulations at exactly the same parameters led to energy-spectra, which can be seen in figure (5.24) and

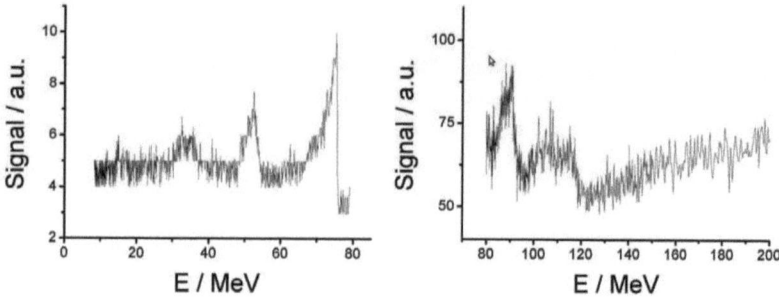

Figure 5.23: *Energy histogram of electrons after experimental bubble acceleration. Data obtained by Ariane Pipahl. Exact data about the electron numbers were not available.*

(5.25) on page 71. Figure (5.24) was done during an early time step of the whole acceleration. It shows a similar multi-peak structure and energies of up to $150 MeV$, although there is another very flat peak at about $220 MeV$. A comparison of the number of electrons trapped in experiment and simulation was not done, due to the lack of information about the number of electrons in the experiments. In contrast to the experiments in my simulations the laser pulse never reached the acceleration length of $2500 \mu m \sim 3000\lambda$ as can be seen in the histogram (5.25). There the multi-peak feature is lost and the energy of the accelerated electrons is smeared out over more than $200 MeV$. Explanations for the factor of 2.5 between the experimental and the simulated acceleration length could be slightly different parameters in the shot presented in figure (5.23). This could include a different plasma density or a slightly higher laser amplitude, which could lead to a larger acceleration length. Also, a shorter gas stream during this specific shot or deviations from the linear gradient and the flat plateau could explain the differences. Still both simulation and experiment show the same multi-peak structure at about the same energies. The reason for this multi-peak structure can be seen in figure (5.26) on page 71. The picture shows a transversal cut through the electron density distribution during the early time in the simulation. There a distorted bubble can be seen and inside the electron bunch shows a clear segmentation. Since the accelerating field is linear to the relative position $\zeta(x)$ (equation (3.32) on page 32) all the different segments are subjected to electric fields of different strength. This leads to several peaks in the electron spectrum presented in figure (5.24) and most likely the experimental histogram 5.23.

In chapter 6 I will present two collaborations with Michael Schnell from the Institut für Optik und Quantenelektronik at the Friedrich-Schiller University Jena. In

5.2. COMPARISON WITH BUBBLE ACCELERATION EXPERIMENT 71

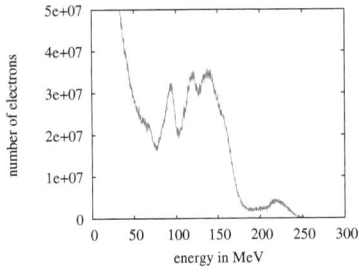
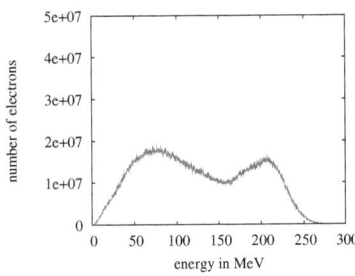

Figure 5.24: *Energy histogram of electrons from simulation at $t = 1000\lambda/c$.*

Figure 5.25: *Energy histogram of electrons after having left the plasma (at $t = 3500\lambda/c$).*

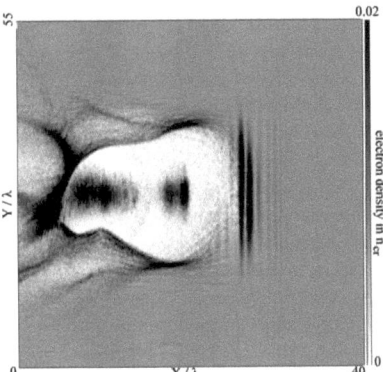

Figure 5.26: *Cut through the electron density distribution of a bubble simulation with parameters corresponding to the experiment done by Ariane Pipahl. A distorted bubble with a strongly segmented bunch of electrons is visible.*

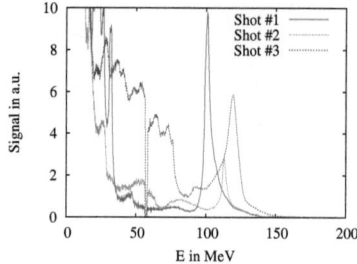 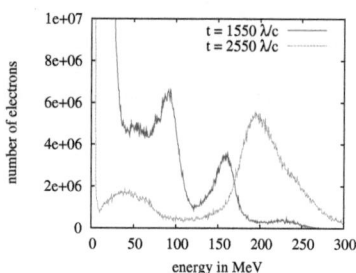

Figure 5.27: *Energy histogram of electrons from three shots during the same experiment conducted by Michael Schnell.*

Figure 5.28: *Energy histogram of electrons after simulated bubble acceleration with the same parameters as in the experiment.*

both works we achieved quiet good agreement between simulation and experiments. Therefore I would like to discuss electron histograms obtained from Schnell's experiments and from my simulations. I was given a sample of typical data obtained from a single acceleration experiment. During this experiment several 'shots' of the laser facility haven been focussed onto the same gas jet in rapid succession. The data from this experiment was used as a basis for a statistical analysis for this experiment. The electron spectrum from three of these shots, which are typical for the whole data set, can be seen in figure (5.27) on page 72. The energy of the trapped electrons from these shots range between $100 MeV$ and $130 MeV$. The peaks are well defined. For all the shots the acceleration length was about $1.3 mm = 1625\lambda$. In figure (5.28) data from a simulation with parameters exactly corresponding to the ones from the experiment is shown. Similar to the simulations of the Pipahl-experiment I found in the simulations higher energies as the ones in the experiments. The histogram at $t = 1550\lambda/c$ fits the experimental results quiet well, especially shot #2 and #3 with only slightly higher energies at about $150 MeV$. After reaching the same acceleration length as in the experiment after a simulation time of $2550\lambda/c$ the electron energies are much higher than observed by Schnell at about $200 MeV$. This implies, that the laser pulse was not depleted after $1.3 mm$ during simulations. Here, only a slight difference in the acceleration length of about 100λ was found.

From this studies, it seems that the VLPL-code is able to simulate realistic scenarios very well including details in field development and particle energies. The only exemption was a significant different acceleration length in one of the experiments. Since this experiment was the only exception, there is not enough data for a systematic analysis of the reason.

5.2.3 Summary

This section dealt with comparisons of experimental data with the PIC-simulation I conducted using the VLPL-code. I focussed on the development of the laser pulse itself and the energy of electrons gained during the acceleration. During the simulations the laser pulse developed very similar to the laser pulse in the experiments. The same red-shift and pulse modulation could be found in both sets of data. However a most likely ion induced, slight blue-shift was absent in the simulation data. This was not surprising since the simulations did not include ionisation. Electron energies and noticeable structures in the electron spectra gained from experiments could be reproduced with simulations. Only slight deviations were found in the spectra. One experimental set of data suggested a very different acceleration length, while other experiments agreed with the acceleration length from the simulations.

Chapter 6

Synchrotron radiation and application

6.1 Introduction

In the introduction I already mentioned, that the creation of radiation is one of the main application for fast electrons. In particular for synchrotron radiation (see chapter 2.3) one needs electrons from very powerful accelerators such as the bubble accelerator. In this chapter specific problems and application from synchrotron radiation generated by electrons within the bubble regime will be presented. Some of the findings of this chapter are the results of collaborations with experimental physicists.

6.1.1 Calculation and representation of radiation and polarisation

In the VLPL-3D code the emission of synchrotron radiation is calculated using the momentum updates which are calculated in PIC-simulations anyway. With the change of the momentum in all spacial directions the code calculates the number, energy and direction of photons being emitted by every macro-particle. The numerical equations are based on the exact equation for classical betatron radiation presented in [1980]. In figure (6.1) on page 76 the synchrotron spectrum during a simulation on a bubble acceleration experiment is presented. The figure shows the spectrum of synchrotron radiation coming from thermal distributed electrons, but with another small peak at several keV. This hump is a result of the quasi mono-energetic electron bunch trapped inside the bubble.

The photons created by the synchrotron module decrease the electrons energy by the amount of energy the photons carry, but otherwise there is no interactions between photons and particles. The photons do not create any fields or interact with particles otherwise. They do not take part in the simulation, but are just stored for

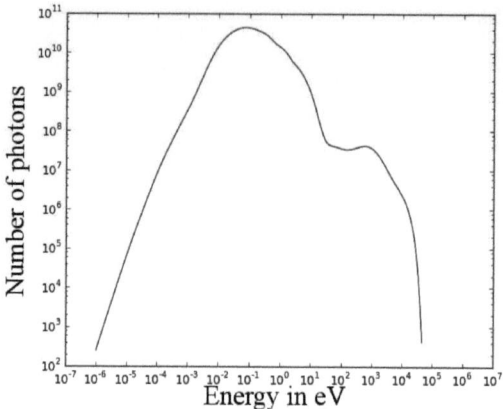

Figure 6.1: *Exemplary synchrotron spectrum during a bubble simulation at $a_0 = 2.1$ and $n_e = 0.00572 n_{cr}$.*

further analysis. The output file for photons contains a three-dimensional histogram of photon energy and the to angles in which the photons have been emitted.

In order to analyse the polarisation of the emitted synchrotron radiation, I included another histogram containing the polarisation of the photons. The polarisation $\mathcal{P}_i$ for a certain direction i is calculated using the absolute value of the momentum change dp_i

$$\mathcal{P}_i^2 = \frac{dp_i^2}{dp_x^2 + dp_y^2 + dp_z^2}. \tag{6.1}$$

If the calculated polarisation only has one non-vanishing component the corresponding photon is linear polarised. The majority of radiation from a bubble simulation although is expected to be a superposition of different polarisation directions and types. Since the emitting electron during a bubble acceleration usually has a change of momentum in forward direction (x−direction) the module also calculates a x−polarisation which in the following will be of no concern. For a reasonable analysis of the transversal polarisation I transformed the equation

$$\mathcal{P}_x^2 + \mathcal{P}_y^2 + \mathcal{P}_z^2 = 1 \tag{6.2}$$

into

$$\tilde{\mathcal{P}}_y^2 + \tilde{\mathcal{P}}_z^2 = 1 \tag{6.3}$$

using the new variable

$$\tilde{\mathcal{P}}_i^2 = \frac{\mathcal{P}_i^2}{1 - \mathcal{P}_x^2}. \tag{6.4}$$

6.1. INTRODUCTION

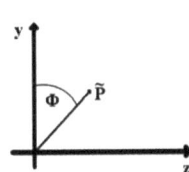

Figure 6.2: *The angle Φ which describes the transversal polarisation of a photon during a Bubble simulation in relation to the transversal spatial axis.*

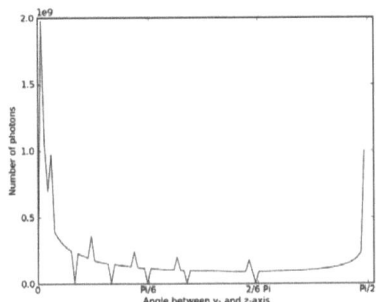

Figure 6.3: *Exemplary plot of the polarisation of synchrotron photons from a Bubble acceleration. The x-axis denotes the angle between the spatial y-axis of the simulation box and the polarisation of a photon, while the $-y$-axis of the plot shows the number of photons with that polarisation direction.*

With this new polarisation variables I like to define the angle Φ as the angle between the transversal axis y and the polarisation vector $\tilde{\mathcal{P}}$ on the 2D-plane spanned by the transversal axis y and z as shown in figure (6.2) on page 77. As been shown in figure (6.2) an angle $\Phi = 0$ means a photon is linear polarised in $y-$ direction, while $\Phi = 45° \hat{=} \pi/4$ corresponds to a photon which is circular polarised in the $z - /y-$plane. This new variable Φ helps visualise the two dimensional, transversal component of the polarisation of photons with a one dimensional variable.

As an example, figure (6.3) on page 77 shows the polarisation of photons from a bubble acceleration. From the predominant $z-$ and $y-$polarisation it is evident that the electrons inside the bubble were oscillating in both transversal direction during the acceleration. The higher amount of $y-$polarised photons is due to the fact that the laser pulse in this simulation was $y-$polarised and some of the electrons already reached the back end of the laser pulse inside the bubble.

6.1.2 Velocity dependent synchrotron radiation

From the findings of Kostyukhov in [2003(Ko)] it can be shown from

$$K = \gamma 2\pi \frac{r_\beta}{\lambda_\beta} \tag{6.5}$$

for $K \gg 1$ and the oscillation amplitude r_β of a single oscillation electron, that

$$r_\beta = \frac{2c\omega_c}{3\omega_\beta^2}\frac{1}{\gamma^2}, \qquad (6.6)$$

with the cut-off frequency ω_c from equation (3.40) on page 34 This means that the transversal oscillation of a electron is quenched during a bubble acceleration as long as that electron gains energy. In [2012] Schnell et al. studied the possibility to deduct the diameter of an electron beam inside a bubble using the emitted synchrotron radiation.

For this work I conducted a series of simulations in order to study the dependence of emitted radiation on the energy of trapped electrons inside a bubble. The set of simulation in the following was done using a linear polarised laser pulse with a Gaussian envelope and an amplitude of $a_0 = 1.5$, a duration of $\tau = 8.11\lambda/c$ and a width of $R = 14.26\lambda$. The electron density was $n_e = 0.006 n_{cr}$. The spatial resolutions were 0.1λ in propagation direction and 0.5λ in transversal direction.

In figure (6.4) on page 79 the energy distribution of all electrons inside the simulation box is shown at different time steps. The red curve corresponds to the energy distribution before quasi mono-energetic acceleration takes place. The green and blue curve show the energy of the trapped electrons as a peak, which shifts during the acceleration to higher energies, while the peak becomes wider. The pink curve represents the electron energy at a time, when the laser pulse used up its energy, the bubble became unstable and there is no quasi mono-energetic acceleration any more. This graph we compared with the synchrotron radiation presented in figure (6.4) on page 79. There it seems, that the majority of photons are emitted during the acceleration of the trapped electrons.

Taking into account more data points we found confirmation of this as shown in figure (6.6) on page 79. There it can clearly be seen that the synchrotron radiation peaks at about the same time as the electron energy. Before and after the main part of the acceleration significantly less radiation is emitted by the whole system. This proved Schnell's claim that most of the radiation detected in his acceleration experiment was originated inside the bubble. Therefore it was legit to use this data to make statements about the shape of the trapped electron bunch.

I also were able to directly confirm the bunch size measured by Schnell (see figure (6.7) on page 80) with direct measurement of the diameter of the electron bunch inside the bubble during simulations (see figure (6.8) on page 80). These results helped to prove that synchrotron radiation from bubble experiments is a valuable instrument for diagnosis of the oscillation of trapped electrons.

One should keep in mind, that bunch size d and the amplitude a of the oscillation of the electrons is not necessary the same. Especially if one considers a situation, in which electrons tend to oscillate in one specific direction y. This can be achieved with the laser polarisation, off-axis injection of electrons into the bubble or asymmetric spot size (see next chapter (6.1.3)). In such a case the amplitude of oscillation in

6.1. INTRODUCTION

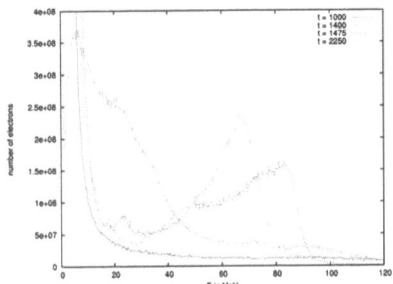

Figure 6.4: *Electron energy histogram of all electrons inside the simulation box. The three different curves belong to the same bubble acceleration simulation at different simulation times.*

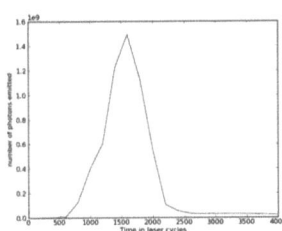

Figure 6.5: *Synchrotron radiation during bubble acceleration. The graph shows the number of photons emitted during a time step of 100 laser cycles.*

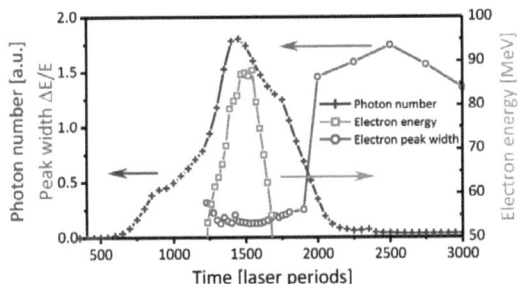

Figure 6.6: *Data about synchrotron radiation from a bubble acceleration in dependants of electron energies and energy spread. Green shows the position of the peak in a electron energy histogram, red the width of the energy peak and blue the number of photons emitted.*

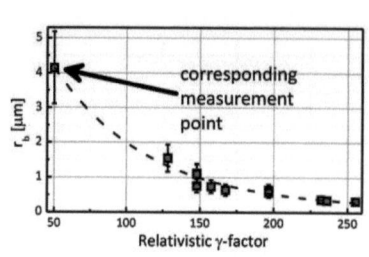

Figure 6.7: *Experimental data of the size of the electron bunch inside the bubble as a function of the electrons average γ-factor as presented in [2012] by Schnell et al.*

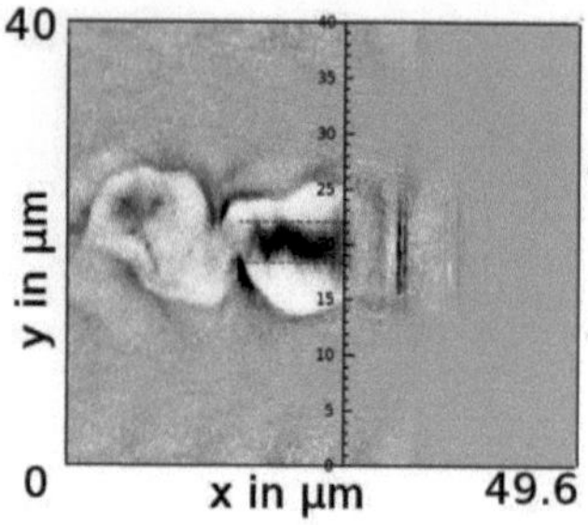

Figure 6.8: *Snapshot of the electron density during a bubble acceleration. This picture is a cut in transversal direction showing the electrons trapped inside the bubble and the diameter of the electron bunch.*

one direction perpendicular to y can be almost vanishing, while the size of the bunch in this direction z does not vanish. Equation (3.38) states, that the electric field of the bubble close to the propagation axis is very low. Yet electrons repulse each other due to their own electric field. Therefore electrons with low initial transversal momentum in z direction and almost on-axis injection into the bubble might travel with only a small oscillation $a_z \leq d_z$ in z−direction. In case of few electrons doing strong oscillations, a and d can be very close to each other. Amplitude and spot diameter are illustrated in figure (6.9) on page 80.

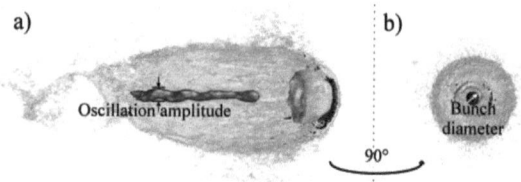

Figure 6.9: *Density distribution of electrons in a bubble acceleration. Blue represents low electron densities, red denotes high densities and the laser spot is illustrated in yellow. a) shows the oscillation amplitude of electrons undergoing betatron oscillation inside the bubble. b) shows the diameter of the trapped electron bunch.*

6.1. INTRODUCTION

6.1.3 Polarisation of synchrotron radiation

As discussed in chapter (2.3) electrons trapped inside the bubble oscillate around the propagation axis. If there are no asymmetries inside the bubble or during the formation of the bubble the electrons should oscillate equally in every transversal direction. As an comparison to figure (6.3) on page 77 which showed the polarisation of an electron bunch which is more or less evenly oscillating in both transversal directions, the pictures (6.12) and (6.13) on page 83 show the polarisation of radiation from unevenly oscillating electrons. In the simulation corresponding to figure 6.12 electrons were created in a already existing bubble, but with a distance of about the bubble radius in $y-$direction away from the propagation axis. Perpendicular to this off-axis direction in the simulation, which lead to (6.13), the electrons were created off-axis in $z-$direction. As can clearly be seen the polarisation of the synchrotron radiation is predominant in the same direction as the electrons have been created in.

With the addition of the calculation and output of polarisation of emitted synchrotron radiation I joined Schnell's research of the polarisation of radiation from a bubble acceleration in [2013]. This study proved that it is possible to control the polarisation of the radiation emitted from the electrons inside a bubble. This was done by either using a tilted pulse front or using a asymmetric laser pulse. For this study I conducted several simulations in order to investigate the dependence of the polarisation of the radiation on the exact shape of the laser pulse. The two modification of shapes I used were laser pulses which were prolonged in one transversal direction and laser pulses with tilted pulse front.

In case of the tilted laser pulse, the longitudinal coordinate x in the Gaussian profile of the laser envelope was transformed into

$$x' = x - y \tan\left(\frac{\pi\psi}{180}\right) \qquad (6.7)$$

in order to obtain a pulse which has a pulse front tilted by the angle ψ in the transversal y-direction (see figure (6.10) on page 82).

In figure (6.14) on page 83 the quotient of all photons with at least 95% $y-$polarisation or $\Phi \leq 0.05 \cdot \pi/2$ over the number of photons with at least 95% $z-$polarisation or $\Phi \geq 0.95 \cdot \pi/2$ is shown. This figure proves that the number of photons polarised in tilt-direction compared to the photons polarised perpendicular to that direction can be increased with tilt angles ψ up to 20°. At higher angles the ratio decreases until at about 40° when the effect is nullified. Comparing this to the experimental data from Schnell we were able to achieve a very good agreement as shown in figure (6.15) on page 83. In this figure again a quotient is shown, but this time of the number of electrons polarised perpendicular to the tilt direction over the number of photons polarised in pulse direction. Since the pulse angles during the experiments were

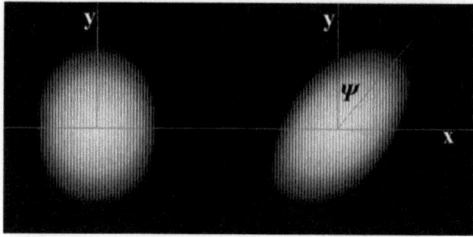
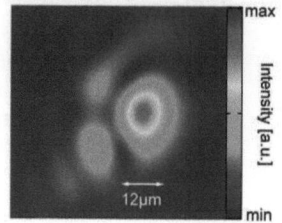

Figure 6.10: *2D-cut of intensity distribution of two pulses. Both are propagating in x−direction. The right one has a pulse front tilted by $\Psi = 30°$.*

Figure 6.11: *Intensity plot of an asymmetric laser pulse used in a Bubble acceleration experiment.*

measured at different points than in my simulations, the tilt angles presented so far are not the same as the ones in Schnell's experiments and had to be transformed in order to obtain the results.

In order to model a asymmetric laser pulse as it was used in several experiments I initialised a second laser pulse with just 10% of the power of the main pulse. Both pulses were close to each other with a distance of 21λ between the center of their focusses. The second pulse was initialised off-axis in y−direction while the main pulse travelled on-axis. The exact shape was modelled according to the measurements of Schnell (shown in figure (6.11) on page 82) during one of his experiments. This pulse shape proved to be the most successful one in order to control the polarisation of the emitted synchrotron radiation. The figures (6.16) and (6.17) on page 84 show a strong prominence in the y−polarisation of the radiation up to about nine times the number of photons than in z−direction. With this results it is evident, that the polarisation of synchrotron radiation can be controlled both with a pulse front tilt and an asymmetric laser pulse. The better results though were achieved using an asymmetric pulse.

6.1.4 Summary

In this chapter I focussed on the synchrotron radiation emitted from the electrons inside the bubble and the polarisation of that radiation. I wrote a routine in order to calculate the polarisation of synchrotron radiation and compared my results with experimental data. My data confirmed that synchrotron radiation during a bubble acceleration mainly is emitted by the trapped electrons. We also found good agreement of the specific frequency of the synchrotron radiation from which the diameter of the bunch of trapped electrons can be calculated. This calculated diameter could be matched with direct measurements of the electron bunch from my simulations. Furthermore we could verify that the polarisation of the synchrotron radiation is

6.1. INTRODUCTION

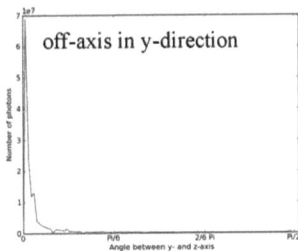

Figure 6.12: *Polarisation of synchrotron radiation of particles injected off-axis in y−direction into the bubble.*

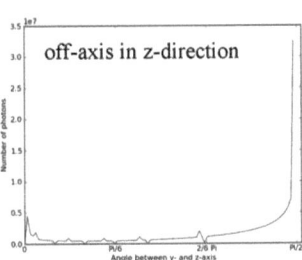

Figure 6.13: *Polarisation of synchrotron radiation of particles injected off-axis in z−direction into the bubble.*

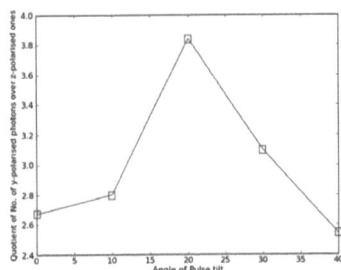

Figure 6.14: *Relative y−polarisation of synchrotron radiation from simulations with different tilted laser pulses. The x−axis denotes the pulse tilt angle in the simulation, while the y−axis shows the quotient of the number of photons emitted in pulse tilt direction over the number of photons polarised perpendicular.*

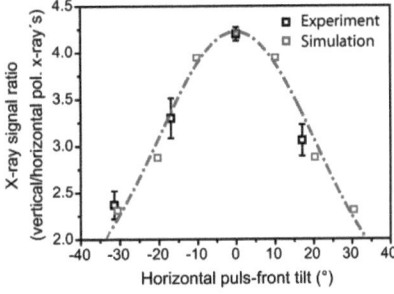

Figure 6.15: *Comparison of simulation data and experimental results. The x−axis denotes the pulse tilt angle in the experimental set-up, while the y−axis shows the inverse of the Quotient from figure (6.14).*

84 CHAPTER 6. SYNCHROTRON RADIATION AND APPLICATION

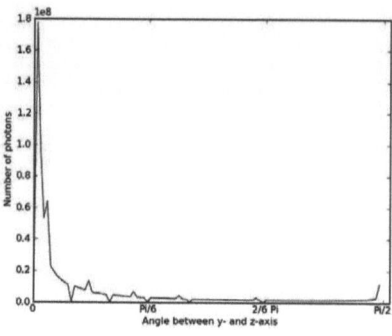

Figure 6.16: *Polarisation of radiation from a bubble created by a laser pulse prolonged in y−direction.*

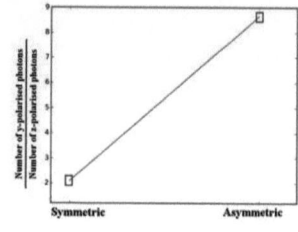

Figure 6.17: *Relative y−polarisation of synchrotron radiation from simulations a symmetric laser pulse and an asymmetric laser pulse.*

directly correlated to the oscillation direction of the electrons inside the bubble. This direction could be controlled in experiments as well as in simulation by using asymmetric or oblique laser pulses. This information made clear that synchrotron radiation is a useful diagnostic tool for bubble experiments.

Chapter 7

Towards advanced bubble acceleration schemes

7.1 Introduction

So far, experiments were only able to accelerate electrons to several $100 MeV$ using the bubble regime. Higher energies were achieved for instance by Leemans et al. in [2006, Jul], but in a slightly different regime using beam guiding. According to scaling laws, energies of even GeV would be possible given strong enough lasers. Unfortunately, achieving these energies require very strong laser pulses, which at the time, this thesis was written, are not largely available. The reason for the need of this high energies are de-phasing and pulse depletion, as explained in the chapters (3.2.3) and (3.3). Since the energy of the laser pulse is depleted at some point and electrons start catching up with the laser pulse, bubble acceleration has a limited acceleration length.

Another problem is the beam quality at high energies. As pictured in figure (5.16) on page 64), the energy spread of the electrons increases at high laser amplitudes due to synchrotron radiation. Therefore the bunch of accelerated electrons has a wider range of velocities and would be less useful for example the application in a FEL than electrons created with smaller laser amplitudes.
Both problems are focus of current research. Two projects I would like to present here.

7.1.1 Side injection

Electron motion during wave breaking is a rather chaotic. Because of this, initial velocities and and positions of electrons at the back side of the bubble tend to vary. Also, self-injection seems to be a process which does take place over a certain period of time. As can be seen in figure (6.9) on page 80, electrons inside the bubble are spread over a certain area and, because of this, are subject to different field

86 CHAPTER 7. TOWARDS ADVANCED BUBBLE ACCELERATION SCHEMES

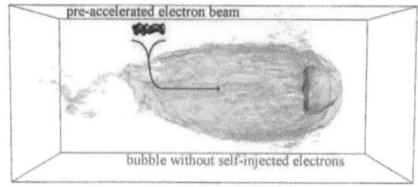

Figure 7.1: *Bubble without self injected electrons (blue) with driving laser pulse (yellow). Electron bunch (red) is injected into the bubble.*

strengths. This of course leads to a certain spread in the energy histogram around the energy predicted by the theory.

In order to reduce this energy spread, one possible way might be trying to avoid self injection. Instead of self-injected electrons with their issues mentioned before, one could try to inject an electron beam with good beam quality into the bubble (figure (7.1) on page 86). From the technical point of view, it is rather simple at low energies (keV or few MeV) to generate an electron beam with a large number of electrons and a small energy spread. If it would be possible to inject such a beam into a bubble, the resulting beam after the acceleration could have significantly improved properties over self-injected electron beams. However, slow, self-injected electrons would interfere with the faster injected electron beam. In order to keep slow electrons from affecting the acceleration by beam loading and other effects, self-injection should be avoided in this scheme.

At the time, this thesis was written, to the best of my knowledge, no reliable information were available about avoiding self-injection, without avoiding the creation of a stable bubble as well. Since the injection seems to be correlated to gradients in the background electron density, I conducted a series of simulations in order to find, whether or not, self-injection can be avoided. Good results were achieved using a slight gradient instead of a density plateau as shown in figure (7.2) on page 87. Figure (7.3) on page 87 shows a cut through the density profile of a bubble along the propagation axis. This bubble, seen in a), lacks the typical electron bunch or stem, which is clearly visible in other figures (for example figure (3.6) on page 36). Still, injection took place, but on a smaller scale. In b) a bunch of injected electrons, performing betatron oscillations, is visible. During this simulation 90% of the injected electrons were trapped inside the bubble. The energy of the trapped electrons in comparison can be seen in figure (7.4) on page 88. The trapped electrons were

7.1. INTRODUCTION

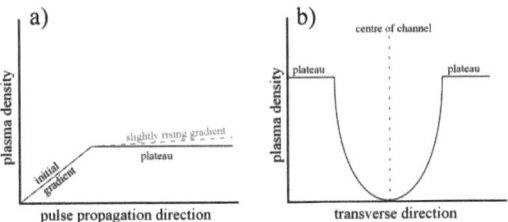

Figure 7.2: *a) Density profile along the propagation direction of the laser pulse for the plateau profile (black) and the slightly rising profile (red). b) density distribution for the channel profile in transverse direction.*

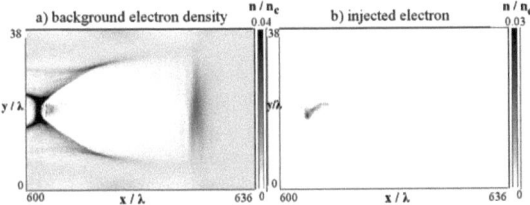

Figure 7.3: *Density profile of a cut through the longitudinal center of the simulation box. The bubble (visible in a) travels in positive x-direction. b) shows the density of side injected electrons inside the bubble.*

accelerated in a mono-energetic fashion with only a few background electrons accelerated at all. However the injected electrons only reach energies similar to those of the background electrons. In this specific scenario the injection of pre-accelerated electrons did not result in higher energies than a common bubble acceleration.

So far I could not find any scalability of this method. Finding the specific values for the length of the initial gradient, the rate at which it is rising, the density, at which the initial gradient stops and the rate of the slightly rising gradient (compare figure (7.2) on page 87) is a lengthy progress. The dependency of these gradient parameters on the spot size or the field amplitude are unclear at the moment. This means, that until now a set of gradient parameters, which actually does prevent self-injection for a given laser pulse, can not be used in order to calculate parameters for different pulse. One reason for this is the fact, that I only obtained working sets of gradient parameters for a very small number of pulses.

88 CHAPTER 7. TOWARDS ADVANCED BUBBLE ACCELERATION SCHEMES

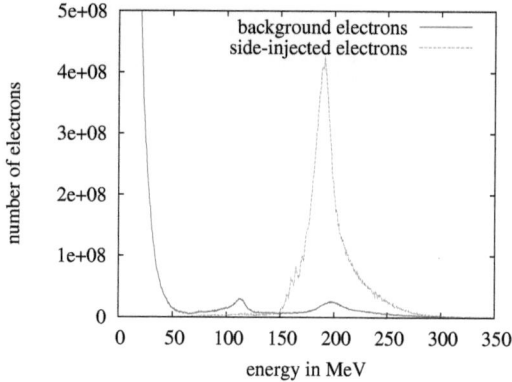

Figure 7.4: *Energy histogram of background electrons (red) and side-injected electrons (green).*

Since the work with different gradients was only partially successful so far, I investigated another option. For this I used a plasma with a channel in longitudinal direction in size of the laser pulse. Such a channel would help focussing the laser pulse over a long distance as shown by Geddes et al. in [2004, May]. It also might counter-act the self-injection because of the small number of electrons on-axis. Here a laser pulse with spot size $R = 6\lambda$ and laser amplitude $a_0 = 6$ was used for the simulations. The density profile used in the following simulations is shown in figure (7.2) b) on page 87. Figure (7.6) b) shows that, unlike in the previous simulation without a channel, almost no electrons from the background were trapped. After an acceleration length of about 800λ the laser pulse started to significantly change shape and the beam quality worsened. An energy histogram of the electrons at an acceleration length of 800λ can be seen in figure (7.6) a) on page 90. There it becomes obvious, that next to no background electrons were trapped. The energy spread of the injected electrons is less than 10%, but the overall energy is quiet low. According to equation (5.11) on page 55 an energy of $216 MeV$ was expected. The simulation ended with an energy of about $150 MeV$ of the electron bunch, which had an initial energy of less than $4 MeV$. The reason for this difference in energies might be the lower density on-axis inside the channel leading to lower electric fields inside the bubble. This is not entirely investigated, yet, but still the result goes to show, that the concept is able to create bubbles without self-injection. Another important factor for avoiding side injection is the pulse length. So far I always tried to use spherical pulses for bubble accelerations. This was done in order to maximise the energy of the laser pulse and use less restrictive parameter for experiments to come. However short pulses proved to be much more efficient in creating more stable

7.1. INTRODUCTION

bubble with less self-injection. This is illustrated in figure (7.5) on page 90. There cuts through the electron density distribution are shown along the propagation axis for two simulations at different time steps. Both simulations feature a laser pulse with spot size of $R = 4\lambda$ an amplitude of $a_0 = 15$ and a channel of width 4λ in the background plasma distribution with a plateau at a density of $n = 0.34 n_{cr}$ according to the scaling laws. In case of a) the pulse duration $0.5\lambda/c$ was used. This resulted in a bubble which hardly changes shape after its formation and with next to no injection. In b) a pulse with duration of $\tau = 2\lambda/c$ was used. The pulse still is shorter than the bubble, which is about as long as wide, but both width and length of the bubble change during the acceleration. Furthermore, the movement of the back relative to the laser pulse injects electrons whenever the back of the bubble falls behind. It seems to be save to assume, that the plasma length has to be longer than the overall laser pulse in order to keep the back of the bubble from oscillating. Since the plasma length is proportional to the optimal spot size,

$$R \sim \frac{1}{\sqrt{S}} \sim \frac{1}{\sqrt{n_e}} \sim \frac{1}{\omega_p} \sim k_p, \qquad (7.1)$$

it seems that the maximum of the pulse length able to avoid self-injection τ_m increases with optimal spot size R. In another simulation with pulse width $R = 6\lambda$, amplitude $a_0 = 6$ and therefore plasma density $n = 0.006 n_{cr}$ a pulse duration of $\tau = 3\lambda/c$ led to positive results as been presented in figure (7.6) on page 90. a) shows an energy histogram at the end of the simulation after 1000 laser cycles. The green curve denotes the injected electrons with energies up to about $150 MeV$, while red shows, that significantly less background electrons were trapped. In b) a cut through the density distribution around the bubble is presented with the injected electrons in green. The back of the bubble is well pronounced and seems quiet stable. These two simulation already suggest, that the maximum pulse duration τ_m is not simply proportional to k_p and therefore to R, since the simulation at $R_1 = 6\lambda$ and $\tau_1 = 3\lambda/c$ would suggest success at $R_2 = 4\lambda$ and $\tau_2 = 2\lambda/c$, instead of the much more restrictive $R_2 = 4\lambda$ and $\tau_2 = 0.5\lambda/c$. From these few simulation it seems, that the mechanism of avoiding oscillation of the back of the bubble and therefore self-injection is not trivial.

7.1.2 Staging acceleration

Using the idea of injecting accelerated electrons into an existing bubble, the next reasonable step seems to be *staging acceleration*. Depletion of a laser pulse as well as de-phasing (see chapter 3.2.3), limit the acceleration length of trapped electrons. Together with the acceleration length, the total energy gain is limited. In order to work around depletion and de-phasing one possible solution might be to use several bubbles instead of just one. In one stage, the electrons of one bubble, after travelling their acceleration length, are extracted from the old bubble. In the next stage, the electrons are injected into a new bubble. One possible way of realising staging

90 CHAPTER 7. TOWARDS ADVANCED BUBBLE ACCELERATION SCHEMES

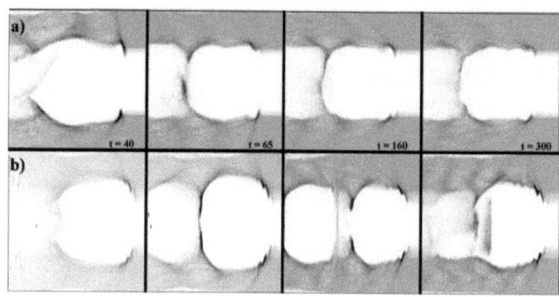

Figure 7.5: *Formation of two bubbles over time. a) using a pulse with a pulse duration of $\tau = 0.5\lambda/c$ and b) with $\tau = 2\lambda/c$.*

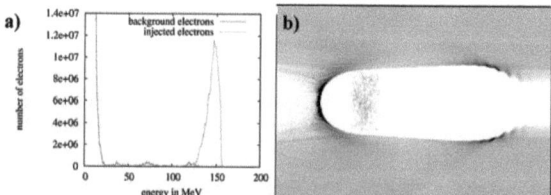

Figure 7.6: *a) Histogram of electron energies at the end of bubble simulation $L_{acc} = 900\lambda$ with spot size of $R = 6\lambda$ and $a_0 = 6$. Red denotes background electrons, while green shows injected electrons in the histogram as well as in the density profile b). b) also shows a cut through the bubble with almost no visible self injection. Side injected electrons are shown in green.*

7.1. INTRODUCTION

acceleration would be to use oblique mirrors in order to separate each stage. A highly over-dense plasma would reflect the laser pulse driving the bubble. Electrons of high enough energy would have a very small cross-section for interaction with the mirror and therefore pass it with little to no effect. The method is illustrated in figure (7.7) on page 93.

Differences to the previous chapter in the method of injection are apparent as the electrons are not accelerated from the side into an existing bubble. Still the tasks of creating a bubble with next to no self injection and the timing of injecting the accelerated electrons into this bubble remain. First, I would like to discuss a simplification of the scheme given in figure (7.7). The simplification was done in order to reduce computation time vastly and divide the problem into two smaller ones. During a simulation of laser-plasma-interaction, the grid resolution in propagation direction of a laser pulse has to be rather high. In most simulations presented to far, the resolution in forward direction was as high as 0.1λ or 0.03λ, while the transversal resolution was much lower (usually 0.5λ). This not only means, that the number of cells in forward direction is much larger, but also is the time step for every PIC-step smaller, as can be seen in equation (4.12) on page 45. A smaller time step means, that more steps have to be computed in order to simulate an acceleration over a given time period. In case of a simulation according to figure (7.7), a laser pulse would propagate on two different axes. One parallel to the overall acceleration axis and one perpendicular to it. This means, that the resolution in this second direction has to be increased as well, increasing the computational cost significantly. Therefore I first simulated a simplified version of figure (7.7), which is given in figure (7.8). This scheme is similar to the one from figure (7.7), with the main difference, that the mirrors are not oblique, but perpendicular to the forward axis. Laser pulses are not created off-axis at the border of the simulation box and re-directed by the mirrors. Instead pulses are created just behind the mirror at the right moment and position as to trap the accelerated electrons passing the mirror.

One of the first simulation to successfully transfer an electron bunch from one bubble (stage 2) into another one (stage 3) is shown in figure (7.9) on page 94. It is important to point out, that not only the parameter already discussed are important (dimensions of the laser pulse and plasma density distribution), but also certain timing related values. Equation (3.32) on page 32 states, that the electric field in forward direction vanishes at the centre of the bubble and then changes its sign, becoming decelerating. Usually during a bubble acceleration, the de-phasing length and the depletion length are the same (see chapter 3.2.3), leading to only one length, the acceleration length (3.44) on page 35. Here, the depletion length is much larger, because of the guiding channel. The laser pulse needs less energy to push aside background electrons. Therefore, electrons from inside the bubble have to be extracted, before de-phasing starts and the electrons enter the decelerating phase of the electric field inside the bubble. In figure (7.9) at the time $t = 200\lambda/c$ the trapped electron bunch already is very close to the centre of the bubble. Extraction therefore is due. The extracted bunch without accelerating field would loose energy travelling through the background plasma. For this reason a new bubble has to be created quickly after the extraction. An important factor for the re-injection is the position, the electrons will have, after the bubble is formed. In the picture at $t = 350\lambda/c$ it can be seen, that the electrons again are close to the centre of the bubble. This is the case just after 60 laser periods after the creation of the new bubble. So it can be assumed that the electrons were injected too far away from the back of the bubble. However an attempt to inject the electrons too close to the back would result in a loss of many of the electrons together with the background electrons, which fail to be self-injected into the bubble. The simulations so far only showed one stage, namely the injection of pre-accelerated electrons into a newly formed bubble. Obviously, further stages with more than one bubble are of interest. The results of a first successful multi-stage simulation can be seen in figure (7.10) on page 94. The red line shows the energy of the electrons in stage 1 at initialisation. After initialisation the electron bunch in injected into the first bubble in stage 2 (green) and from there into two additional ones during stage 3 (blue) and 4 (pink). The final energy of more than $500 MeV$ was achieved after an acceleration of about 1200λ. A regular bubble acceleration should achieve $240 MeV$. However during all stages after the second one, the beam quality deteriorated progressively. To this point it is not clear, whether this can be solved by further optimisation or whether it is a feature of staging acceleration. Also, it seems that during each further stage less and less energy was gained. Since in all cases similar pulses with $R = 4\lambda$ and $a_0 = 15$ have been used, the total possible energy gain using staging acceleration might be limited, unless more powerful pulses are used in later stages. A comparison of the energy from the staging acceleration simulation with electrons without staging acceleration can be seen in figure (7.11) on page 95. The electrons from the staging acceleration (red) clearly show higher energies than the pre-accelerated electrons without further staging acceleration (green) or background electrons (blue) trapped during the channel simulation without staging acceleration. Here again, after the staging acceleration the beam quality was significantly worse than the one without staging acceleration.

7.1.3 Summary

Side injection of electrons has been introduced together with the idea of avoiding self-injection. After evaluating the influence of channel guided laser pulses and very small pulses on self-injection, first successful results of pre-accelerated electrons injected into an existing bubbles were presented. These inserted electrons were accelerated to even higher energies inside the bubble. As one possible application of side injection, the concept of staging acceleration was presented. In this scheme electrons from one bubble acceleration were extracted from that bubble and injected into another one for further acceleration. After a possible, experimental set-up was discussed, a more simplified model was simulated. First successful results of a multi-stage simulation were presented. There it could be seen, that higher energies than with just pre-

7.1. INTRODUCTION

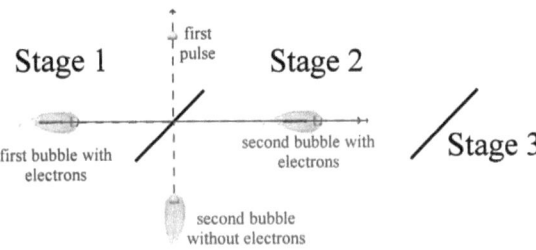

Figure 7.7: *Scheme of staging bubble acceleration. After one bubble acceleration (stage 1) the bubble hits an oblique mirror (black) and its pulse is reflected (blue line). The electrons continue to travel along the black line and are captured by a second bubble travelling along the red line.*

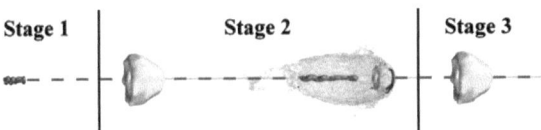

Figure 7.8: *Simplified scheme of staging bubble acceleration. All times are given in λ/c. In stage 1 an accelerated electron bunch travels through a mirror (black) along the red line. A laser pulse (yellow) is created in stage 2 and creates a bubble in which the bunch is trapped. Bubble, electrons and pulse travel until they hit another mirror, which is passed only by the electrons. In stage 3 another pulse is created.*

94 CHAPTER 7. TOWARDS ADVANCED BUBBLE ACCELERATION SCHEMES

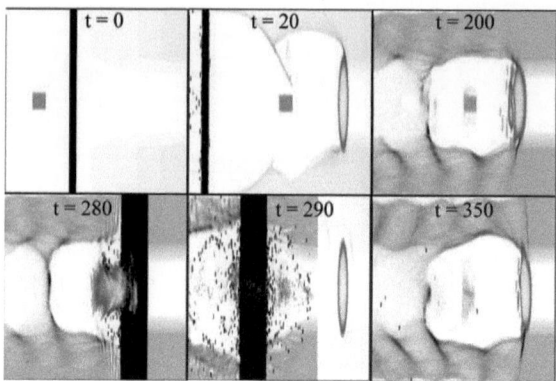

Figure 7.9: *Snapshots from a simplified staging acceleration simulation. $t = 0$ shows the set-up with an pre-accelerated electron bunch (green) and the first mirror (black) in stage 1. At $t = 20$ the electron bunch enters the plasma in stage 2, while a laser pulse (yellow and red) is generated. The bunch is then captured and accelerated inside a bubble ($t = 200$). The accelerating pulse, after almost being depleted, hits a wall at $t = 280$. The bunch travels through the wall into stage 3, where a new laser pulse is created. A new bubble is formed and the acceleration continues ($t = 350$).*

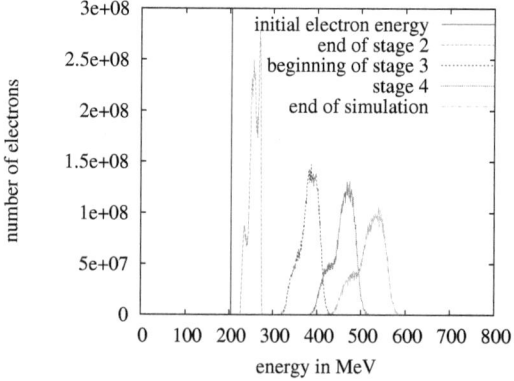

Figure 7.10: *Energy histogram of a simplified staging acceleration simulation (figure (7.8)). Red shows the energy of the electron bunch before injection (stage 1). The other graphs show the progression of the electron beam through three further stages.*

7.1. INTRODUCTION

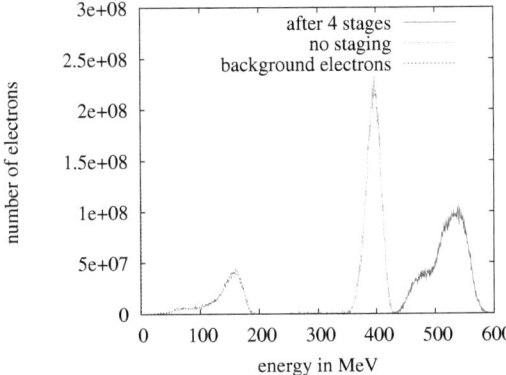

Figure 7.11: *Comparison of energy histograms of electrons after staging acceleration simulation (red), pre-accelerated electrons without staging acceleration (green) and captured background electrons (blue).*

accelerated electrons can be achieved, but with a loss of beam quality. Whether this is can be avoided through further optimisation or if it is a feature of staging acceleration is unknown, yet.

Chapter 8
Summary

The main focus of this work was the bubble regime of laser-plasma acceleration. I investigated scaling laws based on a theoretical model for the bubble regime with numerical simulations. Since the scaling laws were derived utilising dimensional analysis, dimensionless factors for pulse spot size, electron energy and number of accelerated electrons had to be estimated. For this I observed the laser pulse evolution during simulations, calculated its Full-Width-at-Half-Maximum in order to find the exact conditions for a stable laser pulse. I compared electron numbers and energies from simulations to theoretical predictions by conducting a parameter scan. During this scan I found a wide range of parameters in which the theoretical prediction matched the simulation very well. This scan also allowed me to fit theoretical predictions to simulation results in order to obtain the missing dimensionless factors. Comparing the achieved energies of electrons with the predictions from the theory, while also investigating the general shape of the plasma oscillations created at different parameters, I found the lower limits of the parameter space at which simulation and prediction stopped matching each other. At very low spot sizes $R \leq 2\lambda$ or very low laser amplitudes $a_0 \leq 4$, bubbles were not created clearly and observed energies did not match the predictions. After including radiation in the simulations reaction in the form of synchrotron radiation as well as creation of gamma quanta, a upper limit for the laser amplitude at about $a_0 = 128$ became apparent. At this point radiation losses significantly reduce the energy electrons gain during the acceleration, as well as the quality of the electron beam. An expression describing the loss of energy was approximated numerically.

My work included several collaborations, which made it possible to compare results from simulation to data gained from experimental studies. A comparison of simulation data to experimental data form the Arcturus Facility at the Heinrich-Heine-University Düsseldorf proved that, in both simulation and experiment, high densities in front of a laser pulse can lead to pulse modulation and to red shift. Furthermore, I was able to confirm that simulations with the VLPL-code are able to reproduce electron energies very similar to ones obtained from data of corresponding experiments. The simulated electrons achieved slightly more energy, but their histograms showed comparable structures. An anomaly was found in the acceleration length of the simulation, which differed significantly from data frome one experimental campaign. Since other experiments seem to agree with the simulated acceleration length, a specific reason could not be found.

Another part of my collaboration with experimental physicists included a project dealing with synchrotron radiation emitted from a bubble experiment. During this collaboration it was possible to prove that the transverse size of the electron bunch and therefore the amplitude of the oscillation of the electrons inside the bubble can be calculated from the radiation emitted during a bubble experiment. Also, using the polarisation of the emitted radiation, the exact direction of the betatron oscillation can be determined. This direction could be calculated and controlled in both experiment and simulation by using oblique or asymmetric pulses.

The last part of my work gives an overview on projects, which are still subject of research at the time of writing this thesis. The first project focusses on side injection of pre-accelerated electrons into an existing bubble. A key to this work is to avoid the self-injection of background electrons. A preliminary analysis of how spot size and background plasma density distribution have to be chosen is also given. One important application of side injection, the so-called staging acceleration, is presented in the second project. Staging acceleration is a method of accelerating electrons after the pulse driving a bubble is depleted. In this scheme, electrons are extracted from one bubble and inserted into a new one. First successful results of a staging simulation with several bubbles were shown and compared to simulations without staging. Staging was able to achieve higher energies, but at the cost of higher energy spread.

Chapter 9
Outlook

Several further studies suggest themselves based on the results presented in this thesis. The parameter scan presented in chapter 5.1.2 included spot sizes up to a radius of $R = 12\lambda$. This limit was given by the computational effort of simulating larger spot sizes. With advanced PIC-schemes, such as Lorentz-Boosts, larger spot sizes are possible. It would be reasonable to investigate spot sizes up to $R = 30\lambda$ or $R = 50\lambda$ at least in order to include the whole range of experimental reasonable parameters. In particular deviations of the simulation results from the theoretical predictions at $R = 12\lambda$ raise the question, whether this deviation are results of the lower resolution, which had to be used, or of the larger spot size. Also, the question arises, whether or not the loss of energy of the electrons due to synchrotron radiation can be modelled with a simple correction of the scaling laws. This study would need to include an analysis of the depletion length of the laser pulse in the high energy regime with synchrotron radiation, since the radiation reaction could influence the energy loss of the laser pulse.

So far, a comparison of simulation results with experimental data about the energy gain of a bubble acceleration has only be done with data from a few experiments. A more detailed study, including a comparison of the the number of electrons accelerated, might help to find out something about possible limits of PIC-simulations and how to improve them.

Presented in chapter 7 were the first results in regard of electron side injection and staging of the bubble acceleration scheme. No scaling for either method was published, yet. Open questions also include, whether or not staging deteriorates beam quality and whether side injection can be used in order to improve the quality of a given electron beam.

Bibliography

[1950]   H.Motz
Applications of the Radiation from Fast Electron Beams J. Appl. Phys., 22, 527 (1951)

[1959]   O. Buneman
Dissipation of Currents in Ionized Media
Phys. Rev. 115, 503517 (1959)

[1962]   J. Dawson
OneDimensional Plasma Model
Phys. Fluids 5, 445 (1962)

[1963]   C.L. Longmire
Elementary plasma physics Wiley, New York (1963)

[1975]   B.B. Kadomtsev
Tokamaks and dimensional analysis Sov. J. Plasma Physics 1, 295 (1975)

[1977]   F.C. Chen
Introduction to plasma physics Plenum Press, New York (1977)

[1977]   J.W. Connor, B. Taylor
Scaling laws for plasma confinement Nuclear Fusion 17, 1047 (1977)

[1979]   J.Tajima, T.Dawson
Laser Electron Accelerator
PRL, Vol. 43, 4, pp. 267-269 (1979)

[1980]   L.D. Landau, E.M. Lifshitz
The Classical Theory of Fields - Course of Theoretical Physics Volume 2 Butterworth-Heinemann, 4 edition, January 15 (1980)

[1985, July]   D.Strickland, G. Mourou
Compression of amplified chirped optical pulses Optics Communications, Vol. 56, 3, pp. 219-221 (1985)

[1985]	C.K. Birdsall, A.B. Langdon *Plasma Physics Via Computer Simulation* McGraw-Hill, New York, 0-07-005371-5, reprints 1991, 2000 and 2002
[1987]	P. Sprangle, Cha-Mei Tang, E. Esarey *Relativistic Self-Focussing of Short-Pulse Radiation Beams in Plasmas* IEEE transactions on Plasma Science, Vol 15, 2 (1987)
[1990 L]	P.Luchini, H.Motz *Undulators and Free-electron Lasers* Oxford University Press (1990)
[1963 E]	J. Eichler, H.J. Eichler *Laser, Bauformen, Strahlführung, Anwendungen* Springer, Berlin (1990)
[1991]	C.K. Birdsal, A.B. Langdon *Plasma physics via computer simulations* Adam Hilger, New York (1991)
[1996]	E. Esarey, P. Sprangle, J. Krall, A. Ting *Overview of Plasma-Based Accelerator Concepts* IEEE transactions on Plasma Science, Vol. 24, pp. 252-288 (1996)
[1998]	D. Gordon, K. Tzeng, C. Clayton, A.E. Dangor, V. Malka, K.A. Marsh, A. Modena, W.B. Mori, P. Muggli, Z. Najmudin, D. Neely, C. Danson, C. Joshi *Observation of Electron Energies Beyond the Linear Dephasing Limit from a Laser-Excited Relativistic Plasma Wave* PRL, Vol. 80, 10, pp. 2133-2136 (1998)
[1999]	A. Pukhov *Three-dimensional electromagnetic relativistic particle-in-cell code VLPL (Virtual Laser Plasma Lab)* Journal of Plasma Physics, Vol. 61, pp. 425-433 (1999)
[1999, May]	A.Pukhov *Three-Dimensional Particle-in-Cell Simulations of Relativistic Laser-Plasma Interactions* Lecture Notes, Institute of Laser Engineering Osaka University (1999)
[2002, Feb]	B.A. Shadwick, G.M. Tarkenton, E.H. Esarey, W.P. Leemans *Fluid simulations of intense laser-plasma interactions* IEEE Transactions on Plasma Science,30(1 I) pp. 38-39 (2002)

BIBLIOGRAPHY 103

[2002, Mar]   A. Pukhov, J. Meyer-ter-Vehn
Laser wake field acceleration: the highly non-linear broken-wave regime Applied Physics B, Vol 74, 4-5, pp. 355-361 (2002)

[2002, May]   E. Esarey, B.A. Shadwick, P. Catravas, W.P. Leemans
Synchrotron radiation from electron beams in plasma-focusing channels Physical Review E, Vol. 65, 056505, 5 (2002)

[2002, Dec]   A. Pukhov
Strong field interaction of laser radiation Rep. Prog. Phys, Vol. 66, pp47-101 (2003)

[2003(Ko)]   I. Kostyukhov, S. Kiselev, A. Pukhov
X-ray generation in an ion channel Phys. Plasmas, Vol. 10, 4818 (2003)

[2003(Kr)]   W.L. Kruer
The physics of laser plasma interactions Westview Pr., UK, (2003), ISBN 0-8133-4083-7

[2004, May]   C.G.R. Geddes, C Toth, J. van Tilborg, E. Esarey, C.B. Schroeder, D. Bruhwiler, C. Nieter, J. Cary, W.P. Leemans *High-quality electron beams from a laser wakefield accelerator using plasma-channel guiding* Nature, Vol 431, 538-541 (2004)

[2004, Jun]   I. Kostyukov, A. Pukhov, S. Kiselev
Phenomenological theory of laser-plasma interaction in bubbler̈egime Physics of Plasmas, Vol 46, 12B (2004)

[2004, Jul]   A. Pukhov, S. Gordienko, S. Kiselev, I. Kostyukov
The bubble regime of laser-plasma acceleration: monoenergetic electrons and the scalability Plasma Physics and Controlled Fusion, Vol 46, 12B (2004)

[2004, Oct]   J.P. Verboncoeur
Particle simulation of plasmas: review and advances Plasma Phys. Control. Fusion 47, pp231-260 (2005)

[2005]   S. Gordienko, A. Pukhov
Scalings for ultrarelativistic laser plasmas and quasimonoenergetic electrons Physics of Plasmas, Vol. 12, 4 (2005)

[2006, Apr]   W. Lu, C. Huang, M. Zhou, W.B. Mori, T. Katsouleas
Nonlinear Theory for Relativistic Plasma Wakefields in the Blowout Regime PRL 96, 165002 (2006)

[2006, Jul]   W. P. Leemans, B. Nagler, A. J. Gonsalves, Cs. Toth, K. Nakamura, C. G. R. Geddes, E. Esarey, C. B. Schroeder, S. M. Hooker *GeV electron beams from a centimetre-scale accelerator* Nature Physics 2, 696-699 (2006)

[2006, Oct]   A. Pukhov, S. Gordienko *Bubble regime of wake field acceleration: similarity theory and optimal scalings* Phil. Trans. R. Soc. A, 364, pp. 623-644 (2006)

[2009, Jan]   I. Kostyukov, E. Nerush, E. Nerush, A. Pukhov, V. Seredov *Electron self-injection in multidimensional relativistic plasma wake fields* Physical Review Letters, Vol. 103, 17, p.175003 (2009)

[2009, Aug]   E. Esarey, C.B. Schroeder, W.P. Leemans *Physics of laser-driven plasma-based electron accelertors* Reviews of Modern Physics, Vol. 81, 3, p.1229-1285 (2009)

[2010]   A. Pipahl *Simultane Untersuchung von ultrakurzen Laserpulsen und laserbeschleunigten Elektronen* Doctoral dissertation, Heinrich-Heine-Universität Dsseldorf (2010)

[2012]   M. Schnell, A. Sävert, B. Landgraf, M. Reuter, M. Nicolai, O. Jäckel, C. Peth, T. Thiele, O. Jansen, A. Pukhov, O. Willi, M. Kaluza, C. Spielmann *Deducing the Electron-Beam Diameter in a Laser-Plasma Accelerator Using X-Ray Betatron Radiation* PRL, 108, 075001 (2012)

[2013]   M. Schnell, A. Sävert, I. Uschmann, M. Reuter, M. Nicolai, T. Kämpfer, B. Landgraf, O. Jäckel, O. Jansen, A. Pukhov, M. C. Kaluza, C. Spielmann *Optical control of hard X-ray polarization by electron injection in a laser wakefield accelerator* Nature Comm. 4, 2421 (2013)

Appendix A
Acknowledgement

Since not all of the help, I received during my studies, was peer reviewed and therefore referenced in the bibliography, I would like to mention it here.
First of course I like to thank my advisor Prof. Dr. Alexander Pukhov for the opportunity to work in his group, the time he invested into supporting me, his code, which lead to all of the simulation results presented here. Furthermore, I like to thank him for sending me to conferences and referring me to other physicist. This way he helped me working in several interesting collaboration and he trusted me with handling these projects self-reliantly, not without providing help, whenever I needed it.
I thank Prof. Dr. Dr. Carsten Müller, too, for agreeing to be my co-advisor and for his input on my thesis.

I would like to thank Prof. Dr. (i.R.) Karl-Heinz Spatschek and Prof. Dr. Dr. Carsten Müller again for letting me be part of their institute.
Furthermore I am very thankful to Prof. Dr. Willi and the GRK 1203, especially Ariane Pipahl, for interesting collaborations, exchanges and of course for the financial support by the GRK.
In regard to collaborations I especially have to thank Michael 'Quick Mic' Schnell and the SFB TR 18 for two fruitful and very enjoyable collaborations.
All work at the institute would surely have been much more complicated without Elvira Gröters, who masters the art of administrative paperwork AND is able to fix even the most severe problems with a few phone calls. No simulations can be done without computers and no week of computing can commence without at least one power or hardware failure. Therefore I thank Evgenij Bleile for the fact, that I did not have to do all of the simulations with just pen and paper.
Quite a few people I had the pleasure to meet in this Institute. It was a joy working and joking with each one of them: Tobias Tückmantel, Tongpu Yu, Naveen Kumar, Daniel an der Brügge, Götz Lehmann, Liangliang Ji, Johannes Thomas, John Farmer, Phuc Luu, Mykyta Charednycheck, Roberto Martorelli, Matthias Dellweg, Martin Jansen, Friedrich Schluck, Alexey Snydnikov, Ajit Upadhyay, Min Chen,

Sebastian Münster, Adriane Schelin, Christoph Karle and Eckehard Zügge.

Without moral support and someone looking after my sanity once in a while, at times I would have struggled terribly. Quite a big part of my mental health care was done by my gentle and charming girlfriend Alexandra, for who I am deeply grateful. I like to thank all my other friends for helping me relaxing and for being able to stand my occasional lamentations.

A special thank I like to address to my family, my parents Lilli Rosa and Klaus-Ulrich Jansen and my brother Andreas and Michael Cabrera for being always supporting not only during my doctorate, but through my entire studies.

i want morebooks!

Buy your books fast and straightforward online - at one of world's fastest growing online book stores! Environmentally sound due to Print-on-Demand technologies.

Buy your books online at
www.get-morebooks.com

Kaufen Sie Ihre Bücher schnell und unkompliziert online – auf einer der am schnellsten wachsenden Buchhandelsplattformen weltweit! Dank Print-On-Demand umwelt- und ressourcenschonend produziert.

Bücher schneller online kaufen
www.morebooks.de

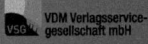

VDM Verlagsservicegesellschaft mbH
Heinrich-Böcking-Str. 6-8
D - 66121 Saarbrücken

Telefon: +49 681 3720 174
Telefax: +49 681 3720 1749

info@vdm-vsg.de
www.vdm-vsg.de

Printed by Books on Demand GmbH, Norderstedt / Germany